AF413746

LETHAL AUTONOMY

THE FUTURE OF WARFARE

WHETHER WE LIKE IT OR NOT

FRANK KENDALL

To all American service members for their willingness to risk and even sacrifice their lives in defense of our country and the values for which it stands.

TABLE OF CONTENTS

PREFACE

This work is a culmination of several decades of studying, working on, and thinking about the intersection of technology and organized violent human conflict. We are living in an age of exponential technology development that is certain to have a profound effect on how any future conflicts will be fought. We are also living in an age in which a brief period of American dominance in conventional warfare may be coming to an end. I do not believe that the implications of those emerging technologies or how they will impact American military power have been adequately explored or defined. This book will certainly not be the final answer to the question of how future wars, especially wars between great powers, will be fought and won, but my hope is that it will stimulate additional analysis (especially quantitative analysis), technology maturation, and experimentation that will point the way forward more clearly.

I have confined this book to the subject of conventional warfare between major economic powers, leaving strategic nuclear conflict, weapons of mass destruction in general, counterinsurgency, and counterterrorism largely to others or another time, although I do address some of those topics briefly. Initially, my intent was to write a somewhat abstract and generic book, but as the work progressed, it became increasingly difficult to avoid taking a US-centric perspective. I finally concluded that the best course was to accept that the problem I was really most interested in was the erosion of US conventional military superiority, something I've been sounding the alarm about for over a decade. Much of what I discuss is generic and could apply to any nation's conventional military, but I have also felt compelled to address the unique situation faced by the US as a great power with major military alliance commitments oceans and continents away.

The US has global interests to protect and potential adversaries who have been actively working to acquire the means to defeat our ability to project conventional military power far from our shores. During the early decades of my career, the United States faced one such problem, the Soviet Union, and one likely theater of war, Europe. Today we face two major threats, China and Russia, and two vastly different theaters, the Western Pacific and Europe. There is arguably an analogy between today and the World War II era when we faced Germany in Europe and Japan in the Pacific, but technology has dramatically changed the implications of time and distance, and the environment in which we confront the strategic threats of this century.

In the US, the Defense Advanced Research Projects Agency (DARPA) has led that technological change, with some more tentative advances taking place in the military services as well. Internationally, the development of and the operational movement toward manufacturing and employment of lethal autonomous systems, and their effectiveness against legacy concepts, has been even more aggressive. The recent conflicts in Armenia, Ukraine, and the Middle East have provided operational demonstrations of some of the relevant technologies.

As I originally completed and then updated this work, I was left with three important impressions that I would like to mention here. They are indicative of critical needs in the human side of the future of warfare. First, while I have emphasized autonomy, in particular lethal autonomy, I have always left an important controlling role for humans in this book. This role manifests itself at an echelon or two above where individual engagements occur. Humans may not do the actual fighting in the future operational concepts that I imagine, but they will be critical to successful outcomes, making their judgment, skills, and preparation a paramount concern.

Second, I envision future conventional conflict to be violent at scale and to involve the full range of adversaries' power to achieve prompt and decisive victory. Our peer or near-peer future adversaries are unlikely to come at us with measured and tentative commitments but with all that they have. Our ability to deter or defeat aggression will depend on our ability to field forces significantly more operationally effective than our enemy's and on our skill in employing them at scale in a violent full-spectrum environment.

Third, our forces today are increasingly obsolete. They represent huge capital investments, and while becoming obsolete, they still are very useful in most contingencies and against our less capable adversaries. It will be hard for our military leaders to move on from them. Nevertheless, whether we like it or not, we are in a race to identify the best path forward and modernize before our peer competitors steal a march on us. This is about humans, their cultures, their willingness to accept risk, and their skill at managing risk. As the Secretary of the Air Force, I had a brief window in which to move the Air Force and Space Force forward. I did my best.

What follows are my thoughts on what the future of conventional warfare might look like. It is not just a technical and operational story; I couldn't resist weaving in some of my own personal experiences over the years where I thought they were relevant. I hope readers will not find these digressions too distracting.

A word on the timeline of this book. I began mentally drafting this book when I left government service in early 2017 and largely completed its first written form before taking office as the Secretary of the Air Force in the summer of 2021. The views within are thus my own and do not represent the official position of the Department of the Air Force or the Department of Defense (DOD).[1] However, I would offer two observations on the continued relevance of this line of thinking.

First, it has been interesting for the last few years to observe the progress of technology, experimentation, and innovative thinking by others

---

1   The Trump administration has used Department of War as a "secondary
     title." This is not legally correct and not used in this book.

all moving rapidly down the general path I will describe here. While I continue to believe that we are "out of time" to implement the concepts herein from a technological, organizational, and doctrinal perspective, I am heartened by the significant movement already underway. Second, the experiences of the last four years have served to reinforce my belief that these concepts are directionally correct. For example, technologies that were demonstrated in Armenia have been further developed by innovative Ukrainians and proven at scale in the conflict with Russia. Unmanned aerial vehicles (UAVs) employed by ISIS as asymmetric threats to US forces in Syria and Iraq have evolved into one-way attack vehicles threatening shipping in the Red Sea. Unmanned surface vessels have devastated the Russian Black Sea Fleet. The cycle of innovation and response only accelerates, and I continue to believe that the appropriate response is to master technological change rather than resist it.

FOREWORD

In *Lethal Autonomy*, Frank Kendall makes a major contribution to our understanding of how warfare is changing and how still emerging technologies are going to change literally everything about how major powers equip, organize, and fight over the next several years. At its heart, *Lethal Autonomy,* is about the emergence of uncrewed weapon systems which will fundamentally transform warfare in all military domains. *Lethal Autonomy* is much more than a summary of the changes we are already seeing. It takes us into the future comprehensively and insightfully unlike anything else on the subject.

I've known Frank Kendall for many years. He knows what he is talking about. He has been a leader in the modernization efforts of the Department of Defense over a long and highly successful career, starting as a West Point graduate and ultimately serving as the Department of Defense lead civilian for all Tactical Warfare Programs and as both the Pentagon's chief weapons buyer during the Obama administration and as the Secretary of the Air Force during the Biden administration. I don't know anyone with broader, deeper, and more extensive experience with military modernization. *Lethal Autonomy* puts all of that expertise on full display.

Perhaps the most impressive feature of *Lethal Autonomy* is its thoroughness and completeness. War is a complex endeavor, made more so by the integration of advanced technologies including automation and artificial intelligence. Frank Kendall appreciates this more than anyone I know. In *Lethal Autonomy* he addresses that complexity head-on. *Lethal Autonomy* begins with a broad discussion of the attributes that fundamentally matter in warfare, how technology has affected those attributes, and how comparative advantage can be measured.

Having set the stage, *Lethal Autonomy* then goes into depth about the operational concepts and key building blocks or weapon systems we can expect to emerge in ground, sea surface, sea subsurface, air, space, and cyber domains. In each domain, Frank Kendall addresses the complexities in operational needs, design requirements, and other military functions specific to that domain. As I read *Lethal Autonomy*, I was hard-pressed to think of any aspect of conventional operations in any domain that Frank didn't address.

The discussion of two domains, space and cyber, are especially noteworthy. There has never been a major conflict in either of these domains. *Lethal Autonomy* provides a compelling, innovative, and clear picture of how these domains could be decisive in a future conflict, and of what the United States needs to do to achieve superiority.

Lethal Autonomy would be incomplete without a thorough discussion of future multi-domain relationships. This is provided in an integrating chapter titled "Joint, Multi-domain, and Combined Operations." In this chapter, *Lethal Autonomy* addresses the significance of the operational dependencies between domains, military services, and allies or partners. These relationships are essential to future operational success. *Lethal Autonomy* provides a focused picture of how the future domain-specific capabilities discussed earlier can be combined to achieve a transformation in warfare.

I particularly enjoyed that *Lethal Autonomy* is a personal story. Throughout this volume, Frank provides anecdotes and references from his personal experience leading the modernization programs of the US military. Frank has always been a hands-on leader. During his tenures as the Under Secretary of Defense for Acquisition, Technology, and Logistics and as the Secretary of the Air Force, he was fully engaged in shaping American military modernization. *Lethal Autonomy* tells us much of that story and makes clear the vision that Frank was trying to fulfill.

The future envisioned in *Lethal Autonomy* is neither obvious and foregone nor science fiction. As imagined by Frank Kendall, it is a practical reality that may well emerge in the next several years. It remains to be seen whether or not the United States will lead this transformation.

**Admiral James Stavridis (ret), former Supreme
Allied Commander at NATO**

CHAPTER 1

SETTING THE STAGE

SOME HISTORY, PERSONAL AND OTHERWISE

I hate war. It is the most wasteful and cruel of human endeavors. And yet I have chosen to spend most of my life either in uniform or working on weapon system development programs as a civilian. I made this choice because I felt that the United States was a leader in the world for things worth fighting for, not just America's narrow interests, but also ideas, values, and principles that were worth defending in their own right. I still feel that way, even if our current political leadership fails to understand that these values, aspirational as they may be at times, are the very core of what makes America great.

I grew up on the shadow of World War II and the Korean War. I started adult life as a cadet at West Point during the Vietnam War. The first twenty years of my career were dedicated to the cause of winning the Cold War against Soviet-style authoritarianism. The last fifteen years of my life were dedicated to the cause of responding to the threat to human freedom posed by new authoritarians in China, and to a lesser extent, Russia. From 2010 to 2017, I served as the Under Secretary of Defense for Acquisition Technology and Logistics or as the Principal Deputy Under Secretary with responsibility for all Department of Defense weapons acquisition programs. From 2021 to 2025, I served as the Secretary of the Air Force, with direct responsibility for leading the US Air Force and US Space Force. During that time, I worked to implement some of the concepts and ideas described in this book. I am hopeful, but not optimistic, that we humans will find another way to resolve disputes between nations than violent confrontation. Until that happens, those of us who live in freedom need to be prepared to defend that freedom.

This book is about the intersection of technology and military operational concepts and organizations. This intersection has been the focal point of most of my career. Technology moves forward at the pace of human discovery and invention. It can be stimulated by the desires and vision of military operators and others, but generally, technology emerges from the endless human search for knowledge which has its own momentum, driven by an array of motivations. Once discovered or created, technology serves as an enabler that drives changes in warfare. It affects the types of force structures or organizations that will be created, the equipment those forces will use, and the operational concepts for how those forces will fight.

The history of warfare (how militaries are organized, equipped, and fight, not the history of battles) is the history of the pursuit of operational advantage, something that changes constantly, and sometimes dramatically, with the emergence and application of new technology. Transformative technologies can render whole classes of equipment and the forces that use them obsolete almost overnight. Ironclads signaled the end of the wooden-ship navies of the world. The longbow, crossbow, and musket signaled the end of mounted armored knights. Airplanes signaled, eventually, the end of the battleship. Precision munitions (predominantly missiles) are currently having an uncertain and still incomplete impact on weapon systems and operational concepts. Looming directly in front of us and emerging more clearly from the experience of the war in Ukraine are the implications of lethal autonomy and artificial intelligence (AI).

My government service spanned a lot of history, particularly two periods of roughly eight years apiece in the acquisition and technology organization of the Office of the Secretary of Defense, separated by fifteen years in the defense industry and then after a four-year break as secretary of the Air Force. The first stint spanned 1986 to 1994—including the last few years of the Cold War and the First Gulf War. The second period spanned 2010 to 2017—a period taken up by conflicts in the Middle East and Afghanistan, but also by increasing awareness of the threat posed by a rising China and a revanchist Russia. I worked hard during this later period to raise the alarm about the military modernization programs being pursued by China, and to a lesser extent by Russia, and to move innovation forward. As Secretary of the Air Force from 2021 to 2025, I was able to direct the course of the Space Force and Air Force.

Because of its resources, regional and global ambitions, and authoritarian system of government, China today is by far the greatest strategic threat to the United States and to the liberal democratic order established by the United States after World War II.[1] As Secretary of the Air Force, my first priority was modernization of the Air Force and Space Force to meet the growing threat posed by China. My experiences have given me the opportunity to participate in endless analysis, discussion, and debate about the best technology investments and the best operational concepts for the United States to adopt in all domains of warfare. It is this combination—technology and operational concepts, applied with thorough preparation and professional execution in relevant organizational structures—that determines success in warfare and defines the evolution of warfare from one era to the next.

At this time, there is a growing awareness of the technologies that will profoundly change warfare in the next decade or two, particularly autonomy and some forms of so-called AI, but with inadequate understanding

---

1 The current Trump administration is challenging the so-called liberal
   world order, but I believe that America's fundamental strategic interests
   are well aligned with those of other democratic nations and that while
   we may not return to the status quo ante of American global dominance,
   the logic of our alliances with other democracies will prevail.

of how those technologies will be operationalized, and without the needed commitment to taking the steps necessary to achieve that goal at scale.

Our great victory in the Cold War came about peacefully as a result of a superior economic system that enabled, among other things, numerous advances in the application of technology to warfare. Some would credit the Reagan-era Strategic Defense Initiative (SDI) for breaking the back of the Soviet economic system. I worked extensively on the SDI program for several years. It was always ambitious to the point of irrationality. While technically and operationally unsuccessful, it may have influenced the collapse of the Soviet Union. More pragmatic, and more demonstrably successful, was the Follow-on Forces Attack (FOFA) program, which grew out of a Defense Advanced Research Projects Agency (DARPA) concept demonstration set of experiments called Assault Breaker. FOFA was a direct response to the Soviet Union's mechanized forces operational concept that depended on successive waves of massed armor. The FOFA suite of wide-area surveillance systems (especially JSTARS, the Joint Surveillance Target Attack Radar System), combined with precision munitions and connected in a supporting data network, plus the introduction of stealth technology, proved in the First Gulf War to be truly revolutionary.

Whatever the motivator or impetus to the Soviet collapse, there was no doubt after 1991 that the US had revolutionized conventional warfare. Unfortunately, we cannot count on our Cold War history to repeat itself against China, our newer and even more challenging strategic competitor.

The quest for military advantage based on the application of technology is a constant in the history of warfare. What makes the last two centuries unique relative to earlier eras is the pace of technological change in general, military and civilian. What makes our current time unique is the transformational effect that specific new commercially based (as opposed to militarily unique) technologies will have on warfare. I believe the changes we will see as a result of the emergence of these commercial technologies are much more fundamental than generally perceived.

One can think of the last century of change in conventional warfare as having come about through the three waves of modernization. The first wave included the industrialization of warfare, dramatic improvements in lethality, and the mechanization of warfare. World War I introduced early versions of new technologies—machine guns, long-range artillery, and barbed wire—that collectively dramatically increased lethality against advancing infantry. The participants in this conflict were roughly evenly matched in capacity to generate forces, and the advantage of defensive firepower over maneuver led to stalemate and a war of attrition—carried out at industrial scale.

As World War I progressed, the widespread use of tanks, trucks, radios, and tactical aircraft set the stage for the culmination of this first wave of modernization that was represented by World War II when those systems restored maneuver to the battlefield. At sea, long-range naval gunfire and heavy armor dominated the conflict originally, but the submarine and the aircraft carrier were introduced and became the platforms that would be dominant in World War II. The introduction of radio communications in World War I and both radar and electronic warfare (EW) were also important elements of this wave of modernization. The operational concept embodied in the "blitzkrieg" concept became the accepted norm during World War II. It was sharpened and reached its culmination in the United States' Air-Land Battle doctrine of the 1980s. I had the privilege as a junior officer to serve under and later work closely with the principal architect of Air-Land Battle, US Army General Donn Starry.

The next wave of modernization is the one the United States introduced in Iraq and Kuwait in 1991 and has used repeatedly since. It remains the underpinning set of capabilities the United States still employs. This suite of capabilities, which originated in the DARPA Assault Breaker demonstration program, appliqués wide-area surveillance, precision munitions—especially standoff missiles—and networked digitized command and control (C2) onto the World War II-era mobile mechanized warfare construct. More accurately, it integrated those technologies onto the Air-Land Battle construct of the Cold War. From 1989 to 1994, I was the Deputy Director of Defense Research and Engineering for Tactical Warfare Programs in the Office of the Secretary of Defense. One of several organizations that reported to me was an office dedicated to FOFA programs that included the main building blocks of this wave of modernization. Included were the airborne radar JSTARS wide-area ground surveillance sensor and several air- and artillery-delivered precision munitions. With strong advocacy from the Defense Science Board (particularly Gene Fubini and Joe Braddock), these programs were managed as a separate portfolio because of their revolutionary and interdependent importance, originally focused on defeating multiple echelons of Soviet armor.

In the thirty-five years between World War II and the First Gulf War, a number of other innovations were embraced, but without as fundamental an impact. Aircraft made the transition from propeller to jet propulsion. Satellite systems came into existence and provided intelligence and communication support. Helicopters came into widespread use for attack, reconnaissance, and maneuver or transport applications. Integrated virtual and live training systems for large ground formations and many-on-many aerial engagements also improved dramatically. Stealthy aircraft were introduced by the United States in the First Gulf War, a significant advance, but not one with fundamental operational concept implications. At sea, the dominance of the aircraft carrier and the submarine appears to have continued, but with significant evolutionary upgrades. None of these innovations changed the fundamental nature of conventional warfare, however.

It can reasonably be said that Dwight Eisenhower, Douglas MacArthur, Chester Nimitz, Hap Arnold, or Alexander Vandegrift would have no trouble understanding the composition of today's American military and how it conducts large-scale conventional operations. That's an oversimplification, but there is a significant element of truth in it. The basic building blocks are highly similar, and where technology has enabled changes, those changes could be understood in operational contexts familiar to the ones employed by these leaders from eighty-plus years ago.

What would most surprise all of those leaders is the relatively low loss rates in combat to which the United States has become accustomed. In the First Gulf War, we expected thousands and even tens of thousands of casualties. There, and in other significant operations since, such as the second assault on Iraq in 2003 and the air campaign in Serbia in 1999, losses of both machines and people for the United States and its partners have been extremely low.[2] This is a historical anomaly, and we cannot expect those results in the future.

After the United States demonstrated the revolutionary efficiency and effectiveness of its conventional forces in Iraq in 1991, all our potential adversaries took notice, but none more so than China. While the United States was entering a period of both recognized military dominance and complacency about the degree to which that dominance would endure, China began to study how it should best respond to the American advances. Coincidentally, China was also starting to become rich and would have both the resources and insight needed to challenge the United States' ability to project power. China has now reached something close to parity with America in gross domestic product and may soon exceed the United States by that measure. When I briefed then-National Security Advisor Susan Rice on Chinese military modernization in 2015, she seemed surprised when I told her I was far more worried about China than Russia. Military power flows from economic power, as do strategic ambitions, and China is vastly more economically powerful than Russia.

To a significant extent, China has emulated the United States as it has worked to modernize its forces. The Chinese have developed and fielded precision munitions, wide-area surveillance systems, and networking. Technologically, China has lagged behind the United States to some degree, but through a combination of theft and intelligent strategic resource application, it has significantly closed the gap in the quality of its fielded capability in traditional warfare domains. In the space domain, China has moved more quickly to field operational capabilities, offensive and defensive, at scale. Most important, China has already developed, or is in the process of developing, a range of capabilities that are well chosen to reduce or eliminate US advantages. These include a suite of capable air-to-air missiles, electronic warfare systems, and anti-satellite systems. What has most dis-

---

2   Counterinsurgency campaigns have continued to extract high numbers of casualties. By their nature they are wars of attrition and put defenders at risk as they are forced to move among the population while trying to suppress insurgent groups and provide security to civilian society.

tinguished China's modernization program, and focused my attention, is the degree to which China has strategically targeted America's ability to project conventional military power close to China.

Unconstrained by the Intermediate-Range Nuclear Forces Treaty, which banned ground-launched conventional as well as nuclear weapons between five hundred and fifty-five hundred kilometers in range for the US and Russia until the US withdrew in 2019, China has emphasized long-range precision conventional missiles, initially both ballistic and cruise missiles, but now hypersonic versions of both as well. The reason for this Chinese emphasis is America's dependency on a small number of high-value assets to conduct power projection. Those assets include aircraft carriers, regional fixed air bases, military satellites, and a few logistics and C2 nodes. China complements its large inventory of ground-based missiles with air- and ship-based systems as well. The short version of the current situation is that the United States is the dominant military power until within about one thousand miles of China, and then things begin to change.[3] That number is growing over time as China fields more capability, but it is already large enough to make it problematic for the United States to meet its security commitment to regional allies.

China, except for hypersonic weapons, is not pursuing a new wave of modernization, per se. It is investing in the same technologies and concepts the United States introduced, just in a way that is tailored to their most important operational military problem—keeping the United States out of what they see as their part of the globe.

The recognition of this problem, which is well beyond "emerging" at this point, led to a number of initiatives over several administrations. Almost immediately after I returned to the Pentagon in March 2010, it became clear to me that we had a big problem. As I reviewed daily intelligence summaries on China's military modernization programs, their strategy became obvious. I started to raise the alarm, but I confronted two problems in getting the Defense Department's attention. The first was that we were still heavily engaged in Afghanistan and Iraq. The second

---

3   This was how I described the situation to then-National Security Advisor Susan Rice in 2015.

was that the idea that the conventional forces of the United States could be challenged, let alone challenged successfully, was an alien and almost unacceptable concept to even articulate. The denial was strong, even when confronted with direct evidence. I continued to beat the drum, internally and externally in speeches, in testimony (both classified and unclassified), and in internal budget deliberations. On more than one occasion, very senior people asked (directed) me to reduce the volume of the alarms I was sounding. By then I was on the record publicly, including in congressional testimony, and I could not do so, even if I had wanted to.

At that time, there were some efforts in the Defense Department to address the problem of China's modernization program and the rising strategic threat China was becoming. When I returned to the Pentagon in 2010, there was already a group called the Long-Term Competitive Strategies Group working on the China problem. It was led by Michèle Flournoy, the Under Secretary of Defense for Policy, and General James "Hoss" Cartwright, Vice Chairman of the Joint Chiefs of Staff. I quickly joined this group and formed a Research, Development, and Acquisition Task Force led by Alan Shaffer, Principal Deputy Assistant Secretary of Defense for Research and Engineering, to support the effort. Later, when we developed a new National Defense Strategy under Secretary of Defense Leon Panetta, it reflected a shift to emphasize the Asia Pacific region. The Obama administration asserted that this shift was not about China, but of course, it obviously was. Secretary Ash Carter later made great-power competition one of the themes of the 2017 defense budget the Obama administration submitted. He emphasized five threats, in this order: China, Russia, North Korea, Iran, and extremist groups. The first Trump administration, in its National Security Strategy and National Defense Strategy, echoed this shift and increased the emphasis on great-power competition as did the Biden administration.[4]

---

4 The second Trump administration has released a new National Security
  Strategy and a new National Defense Strategy that dramatically changes decades
  of successful bipartisan US policy. It emphasizes US homeland defense and
  preeminence in the Western Hemisphere as higher priorities than China or Russia
  and berates our long-standing European allies. This makes no sense to me.

Perhaps the most visible attempt to counter China's modernization program during this period was the so-called Third Offset Strategy. This initiative was created by Deputy Secretary of Defense Bob Work during the final years of the Obama administration. It was based on the idea that the US had successfully implemented two earlier offset strategies and needed a new one. The first of these was the fielding of thousands of low-yield tactical nuclear weapons in Europe to counter the threat of over-whelming formations of massed Soviet armor. After World War II, the US and its allies had largely demobilized and did not have conventional forces large enough to compete with those of the Soviet Union. What the US did have was a short-lived monopoly on nuclear weapons. This first offset strategy was considered successful, until the Soviets started fielding their own nuclear weapons. The second offset strategy was the round of conventional warfare modernization commented on earlier. In this strat-egy, the combination of precision weapons and battlefield awareness was intended to defeat the same massed Soviet armor. The problem Bob Work correctly perceived was that with the proliferation of precision munitions, the second offset strategy would lose its comparative advantage. He was basically right about this; the US is more vulnerable to long-range preci-sion munitions than our adversaries because we project power far from our shores, utilizing forces that depend on small numbers of very capable but expensive and targetable assets.

In the summer of 2015, several of us worked with Bob to define the content of a Third Offset Strategy. That group included Bob Work, Vice Chairman of the Joint Chiefs of Staff Admiral James "Sandy" Winnefeld (and later, General Paul Selva), Defense Science Board Chair Craig Fields, DARPA Director Arati Prabhakar, Assistant Secretary of Defense for Research and Engineering Steve Welby, Acting ASD for Acquisition Jimmy MacStravic, and myself. We met for three or four hours on several occasions to consider options and review information from various sources. During this period, I also initiated a Long-Range Research and Development Program Planning Study led by Steve Welby.

At the end of all this effort, we had come up with three features we thought would be central to a Third Offset Strategy: range, autonomy, and quantity at cost. I still think this is the right formulation, and that it can apply in multiple domains. Our work was incomplete, however, in that we never defined how these principles would be operationalized. What would we actually build and how would those components interact in an operational concept? Without answers to those questions, we were left with the idea of a Third Offset Strategy but no clear definition of what it was and no actionable plan to implement it. This book began as an attempt to move that work forward and to present some more specific ideas about what the next—pick your phrase—Military Technological Revolution, Revolution in Military Affairs, or Offset Strategy will consist of in meaningful operational and technical terms.[5]

The primary way the focus on great-power competition manifested itself in the first Trump administration was an obsession with speed in acquisition programs. That emphasis is continuing in the second Trump administration. I believe this is a serious mistake for two fundamental reasons. The first is that haste makes waste. Trying to do an acquisition program on an arbitrary schedule faster than it can reasonably be completed has historically led to failure in the form of large schedule and cost overruns and nothing entering the inventory. History is full of examples. These include: the Army's Future Combat Systems, the Navy's A-12, the Air Force's Single Integrated Air Picture, and the intelligence community's Future Imagery Architecture. Unfortunately, facts and data never seem to overcome the unfounded optimism of amateurs with what they think is a new idea for how to acquire weapon systems.[6]

---

5   As an aside, at the conclusion of our efforts, Bob Work built a briefing around autonomy alone and started to present it as the result of the Third Offset effort. Bob's product provided a general description of how autonomy and artificial intelligence of various types would transform warfare, but it was abstract and very generic. There was still no clear description of what would be built or how the components would work together in an operational concept. Bob declined to include range and quantity at cost as equally important parameters with autonomy and artificial intelligence. I believe these were important omissions.

6   The illusion that there is some easy way to dramatically shorten acquisition lead times has been the bane of my existence. Throughout my entire multi-decade career, the demand for faster programs has been a constant. It is easy to write down an aggressive or unrealistic schedule; it is hard to deliver a complex weapon

The second is that going fast in the wrong direction doesn't actually get you closer to your goal. We are in a strategic long-term competition with China and arguably with Russia. We need to be smart about what we choose to buy. Speed isn't the only variable we care about. We also normally keep the weapons and other military systems we buy in our inventory for decades. If we expect to continue to do that, then it's important that they be cost-effective and designed for the required service life. If not, then we need to plan and budget for short product life cycles. It's also important that we commit to programs that will actually work in a realistic operational context. The defense industrial ecosystem, traditional and new entrants, has a lot of advocates for innovative ideas, but they haven't always thought their way through all the operational implications of those ideas.

A case in point was the Air Force's multibillion-dollar airborne antimissile laser program. This huge laser mounted in a Boeing 747 could shoot down missiles in boost phase, but the cost of always having these lasers on station and keeping them alive made the system unaffordable and impractical. Unfortunately, this was all discovered several years and several hundred million dollars into the program. The rush to field new technology in immature designs for concepts that don't actually work in practice is not the path to success. Unfortunately, we are currently seeing a lot of that type of behavior from the second Trump administration.[7]

FUNDAMENTALS

For now, let's ignore the notions of offset strategies or revolutions in military affairs and focus more directly on military fundamental attributes and the dominant variables in conventional warfare. I don't want to get bogged down in semantics or taxonomy here, but it's useful to start with a

---

system. I have never encountered anyone who wasn't motivated to deliver products as quickly as possible. If we want products faster, we have to simplify the products we buy. That means accepting less capable products. There is no other way.

7   Golden Dome, the aspirational national missile defense system, is a perfect example.

logical framework so that a reader will understand the origins of some of my thinking once I get into specifics.

All attempts to achieve technology-based military advantage are rooted in some combination of fundamental attributes and variables that affect military success. All major improvements in military performance are also ultimately about some form of greater operational efficiency that has its origins in advantages in superior attributes or relative control over dominant variables. In operations research terminology, this all manifests itself in exchange ratios—kill ratios and cost-exchange ratios. If one side can find a way to better destroy the other side's assets while losing fewer of its own, in both quantity and cost metrics, then that side has an advantage. If this gap is wide and difficult to overcome, the advantage is almost certainly decisive, and may be enduring, at least for a few years and potentially for a decade or two.

When I first wrote the report about the future of warfare a few years ago, we already had the example of achieving high cost-exchange-ratio efficiency in the successful use of inexpensive drones to attack Armenian armor and fixed positions in the conflict between Armenia and Azerbaijan in the summer of 2020.[8] More recently, we have seen the use of drones proliferate in the Ukraine conflict with the same result. For some, the jury is still out on how significant this development really is. I think these examples are clear indicators of where warfare is heading next.

Historically, and for the US, the first two so-called offset strategies provide good examples of fundamental changes in battlefield efficiency, but with very different approaches to achieving operational efficiency. The first offset strategy increased operational efficiency through the ability of a single expensive weapon, a nuclear weapon, to kill a lot of enemy platforms and people. The second offset strategy increased efficiency by using a lot of relatively inexpensive but very accurate individual weapons to each kill a single platform—one shot, one kill. Both modern offset strategies were about achieving cost-effective lethality, but by using the

---

8   John F. Antal, *7 Seconds to Die: A Military Analysis of the Second Nagorno-Karabakh War and the Future of Warfighting* (Philadelphia: Casemate, 2022).

opposite approaches. I expect the next round of change to be about the lethality of relatively inexpensive reusable autonomous systems.

Overall, what attributes and variables drive operational efficiency metrics? The attributes of most direct significance are cost, lethality, and survivability. These, in turn, are all coupled to advantages in range and speed, and in detection and concealment. Every one of these variables or attributes is connected and dynamic, but let's consider each briefly, and consider the United States' current posture with regard to these parameters. All these variables, and many others, must be traded off with each other in design studies and analysis.

My trade-offs in this paper will be primarily intuitive, based on my decades of experience doing, managing, and learning from operational analysis. My original intent was that the report this book is based on would lead to deeper, more analytical trade studies and better-informed decisions about new American weapons programs. As the Secretary of the Air Force, I worked to ensure that operational analysis informed all the decisions we made for modernization of the Air Force and Space Force during my tenure. I took a number of steps to increase the Department of the Air Force's analytical capacity, including elevating the Studies and Analysis Office to be a direct report to the Secretary of the Air Force. The entire Department of Defense still needs to increase its capacity to conduct operational analysis.

Let's start with cost. Cost is usually measured in absolute terms (dollars), but in practice, relative cost is what matters. A good example is using a $100,000 air-to-air missile to shoot down a $1,000 drone as opposed to using the same missile to shoot down a $50 million tactical aircraft. When Iran launched its drone and missile attacks on Israel, some very expensive missiles were used to shoot down relatively low-cost drones. The same thing has happened repeatedly during the Russia-Ukraine conflict. This type of adverse cost-exchange ratio isn't sustainable.

There are some other complicating factors when one analyzes cost in a specific conflict, however. One of those factors is that cost isn't a perfect measure of value. Value assessments are relative and they have a nonlinear quality. Think of it this way: The first and last aircraft carrier in the inventory may have similar costs, but as carriers are lost in combat and can't be replaced, the operational value of the remaining assets increases substantially. The "cost" in operational terms of an incremental loss increases disproportionately and nonlinearly as the asset pool shrinks. Also, if we only have one of something and we lose it, the cost of that loss is 100 percent of that capability. If we can't replace that system for a matter of years, the operational cost of that loss could be very high because of the overall impact it would have.

The US has this problem in spades today because we are very dependent on a small number of assets for critical functions. Independent of their dollar cost or even their cost-exchange ratio, these assets have enormously high value in our operational plans and concepts of operation. Loss of these assets, or a high fraction of them, could lead to operational or even strategic failure very quickly.[9] I believe that China in particular has identified this US weakness and has modernized specifically to take advantage of it.

In addition to cost in dollars, there is also cost in human lives. The US may have a significant and enduring disadvantage when it comes to our asymmetric valuation of human life compared to opponents. To our credit, we are reluctant to expend large numbers of American lives. Our experience for the last few decades has been with operations that have involved comparatively low numbers of casualties. This is a relatively recent development. We had over fifty thousand casualties in Vietnam. We expected over ten thousand casualties in the First Gulf War. We shocked ourselves, in a good way, with the efficiency improvements we had built into our Joint Task Force in Kuwait and Iraq in 1991. We did experience significant losses in Iraq and Afghanistan counterinsurgency campaigns, but not on the scale of previous major conflicts. We bore those losses, but we also took extraordinary measures to reduce them. My guess is that

---

9   My suspicion is that this situation exists today and already limits
     our practical operational options in some real-life scenarios, and
     therefore weakens our conventional deterrent substantially.

if the motivation for the conflict was strong enough, the US could still accept high losses to achieve a vital national military objective, but this proposition hasn't been fully tested for the current generation or for an all-volunteer force.

We also have a cost problem of a related but slightly different nature for another reason within the systems that we have in higher quantities, such as fighter aircraft and surface combatant ships. It may be true that we would not lose a high fraction of these higher-quantity assets in the opening hours or days of a conflict. However, at reasonably foreseeable exchange ratios against a peer competitor, these forces would be fairly quickly attrited. Even modest attrition rates become unacceptable very quickly when measured in the fraction of our capacity that would be consumed in continuous operations over just a few weeks. A good example is the acceptable loss rate for continuous air operations, such as those waged over Germany in World War II or North Vietnam during that conflict. The problem is compound interest in reverse. A loss rate of 1 percent to 2 percent a day in continuous operations may be acceptable on a given day, but over consecutive days it consumes a force very quickly.

In a study I participated in at the Center for Strategic and International Studies (CSIS) led by former Deputy Secretary of Defense Kathleen Hicks (then at CSIS), we reviewed some alternative future force structures that had been hypothesized under different defense budget assumptions. The minor epiphany that struck me was that even the largest of the forces being considered could not sustain anything near historic loss rates in a protracted conflict, either episodically or through continuous attrition. As with the high-value, small-quantity assets discussed above, the replacement time for the systems in our inventories is measured in years. The implications of this situation are strategically significant. We cannot solve this problem with the resources we have if we continue to buy the types of tactical systems we've been buying for decades.

It's worth thinking about our solution to this same problem during the Cold War when we were trying to stop a potential Soviet invasion of West Germany. Even though we had significantly higher fielded inventory and much more force structure than today, we still assumed we would fall back on tactical nuclear weapons if or when the conventional force was about to be overwhelmed. Today we no longer have that disturbing option.[10] There has been some discussion in recent years of acquiring more military production capacity to provide a reserve surge capability for high-priority systems. Having built many defense budgets over the years, I do not see how this could be realistically achieved.

Let's move on to lethality and survivability. These variables are inherently relative and measured in relation to specific threat capabilities. They tend to be thought of on a weapon and platform basis respectively, and that's a good place to start, but force lethality and survivability are also important and bring in added complexity. The armor protection versus anti-armor munitions contest of the Cold War is a good example of the platform versus weapon perspective. This contest was an exercise in continuous incremental competitive improvements in lethality and survivability on both sides. While these attributes can be measured on a system-by-system basis, it can be more important to assess them for forces consisting of combined systems. For example, just the addition of the JSTARS airborne wide-area targeting radar to the First Gulf War battlefield in 1991 dramatically improved the lethality of the US-led coalition force.[11] Lethality and survivability metrics are quantifiable at both levels and lend themselves well to comparative analysis and trade-offs. Former Secretary of

---

10   I'm not recommending here that we rebuild the tactical nuclear forces of the '60s, '70s, and '80s, but we should think carefully about the implications of this situation. I'll leave that problem for another project, but my preliminary thought is that we should aim for a conventional deterrent with the ability to defeat any likely adversary's act of military aggression operationally. This is as opposed to keeping a force in being with the ability to win a protracted conventional conflict against a peer competitor armed with nuclear weapons.

11   I was part of a group that fought successfully to get the two JSTARS developmental prototypes sent to the Gulf, over the objection of the Air Force, which didn't want to disrupt the test program. My favorite quote from that era was by the Chief of Staff of the United States Air Force after the war who said, "We will never go to war without JSTARS again." My own comment as I lobbied for deployment of the two prototypes was "What war are you waiting for?"

Defense Jim Mattis strongly emphasized improving lethality of US forces during his tenure.[12] I'm a huge admirer of Jim Mattis, but I think he got this wrong; US forces have a much bigger problem with survivability than with lethality. I also think there is a strong cultural bias in any military, US or otherwise, currently and historically, to discount vulnerabilities that implicate a military organization's sense of identity. It often takes a shocking operational event to force this recognition, and sometimes even that doesn't do the trick. Examples are legion.[13]

There are other variables to consider. Range, or the ability to act from longer range than the enemy, to both see and kill from farther away, has been at the heart of many, if not most, major advances in warfare over all of human history. Range can confer survivability, but it is only of value in improving operational efficiency if adequate lethality and acceptable survivability and cost are also available with the increased range. Speed, both speed of movement and speed to deliver effects that derives from certain physical features originating at the platform or weapon system level (such as supercruise in the F-22, or most recently hypersonic missile propulsion), has operational value. Speed confers the superior ability to create or shorten range, and in the aggregate, the ability to maneuver forces more rapidly, implying the ability to, in the much-overused phrase, "move to a position of advantage." An ageless adage for fighter pilots is "speed is life," referring to the value of the superior ability to separate from or close with an adversary in aerial combat. Superior speed of movement has military value, but it is very context-dependent and usually does not provide an enduring advantage. Adversaries can emulate major advances in this area relatively quickly.

Another, and neglected, fundamental is sustainability, which I define broadly to encompass everything necessary to bring a force into battle and keep it effective throughout its employment. Professionals do worry

---

12   As has Secretary Pete Hegseth.
13   At the outset of World War II, the commander of US Army cavalry is reported to have written Chief of Staff of the US Army George Marshall to argue that increasing the size of the cavalry branch was the key to winning World War II.

about logistics, and any discussion of the future of warfare must take this into account. It can be great fun to imagine new operational concepts and new platforms, weapons, and force structures, but that's only half the problem of conducting military operations. Forces have to be ready when needed, and they have to get to the battlefield and be supported efficiently and fully once they arrive.

Sustainment considerations are also critical to total life cycle cost. Cost considerations have to take the total cost of operating systems into account when exchange ratios and value comparisons are made. For a given weapon system, sustainment costs are generally much higher than the costs of development and are comparable to or greater than the costs of production. Sustainment is often, if not usually, the most significant contributor to the total lifetime cost of ownership. An enormous part of the cost of our current human operator-centric military enterprise is the cost to operate and maintain the equipment and train the people associated with the equipment.[14] The current US paradigm for many systems, and the nature of modern warfare up to this point, leads to complex systems that require the almost continuous training of human crews on their own organization's equipment to maintain proficiency in very perishable human skills. Automation and the development of AI, and as we shall see, the increasing availability of autonomous systems, is fundamentally changing this part of the military value equation.

For the US in particular, sustainment is part of a vicious cycle. Complex and expensive human-operated systems can only be afforded in small numbers. Because they can only be afforded in small numbers and because they take a long time (years) to manufacture, they also have to be highly survivable, reliable, and field supportable so they can be kept on the battlefield throughout a conflict. All of this drives support cost up. In the US model, our platforms are also used frequently, if not continually, for training purposes, which in turn leads to continuous maintenance at the

---

14   There are multiples of as many people in uniform that are doing support work as there are actual fighters. The Army, for example, has about three support people (including logistics, life support, and headquarters) for every one warfighter.

operational level punctuated by higher-level depot or equivalent maintenance. Any significant upgrades have to be scheduled around the human operators' peacetime operational deployment and training tempos. In wartime use, our complex crewed systems must be maintained at or as close to the operational echelon as possible so they can stay in the fight. A great deal of the current US force structure, at all echelons, is necessary today to maintain relatively small numbers of actual fighting platforms and to provide the trained and deployable people to operate them.

For contrast, it's interesting to compare the US and the Soviet sustainment models. The Soviets designed their systems for their expected battlefield life, not for decades of use in training. The Russian World War II experience was a very high attrition rate over a protracted period. They accepted that high losses in people and equipment would occur. If their World War II experience-based models of combat attrition indicated that a main battle tank would have a three-day life in combat, why design it to be reliable for more than a few days? It would be destroyed by then anyway. In peacetime, the Soviets also only used a small fraction of their equipment in training, and they used the same equipment repeatedly. The balance was parked and only checked occasionally for readiness.[15] As we think about future forces and operational concepts, we should treat sustainment as both a critical parameter and cost factor, and we should be open to new sustainment paradigms that technology might enable.

By now, some readers are likely to be apoplectic that I haven't talked about decision speed and quality. We live in the world of rapidly maturing AI, and the US has long emphasized decision-making superiority and the ability to conduct coordinated action more effectively than an opponent. This attribute of Command, Control, Communications, and Battle Management (C3BM) is a significant contributor to relative efficiency and

---

15   The Ukraine conflict exposed a weakness of the legacy Soviet approach. Russian equipment had not been maintained, and it failed miserably during the initial invasion of Ukraine in 2022.

advantageous exchange ratios.[16] In the United States, this emphasis also dates to the Air-Land Battle doctrine. It was integral to FOFA operational concepts in the 1980s and provided the focus of both digitization and network-centric warfare in the early '90s. Most recently, the quest for decision speed and quality takes the form of Combined Joint All-Domain Command and Control (CJADC2) and in each of the military services' versions of that concept.

This aspirational goal had not been converted into reality, or anywhere near reality, when I took office as the Secretary of the Air Force in 2021. I made a number of organizational, personnel, and management changes designed to move CJADC2 into reality for the Air Force and Space Force. When I left, they were still works in progress, but progress was being made. I view the C3BM function as supporting an operational concept, not the other way around. My view is that, with effort and attention, we can build the C3BM networks we would need to support the operational concepts I will define. C3BM and automated functionality within the networks are important design considerations and enablers, as well as potential vulnerabilities. I'll discuss them later in this book, but they won't be my central focus.

The Soviets were a little obsessed with quantified theoretical measurements of relative combat power, what they called correlation of forces, but they also firmly believed that time was the most important variable on the battlefield. They had a point. Today's advocates of unlimited connectivity and infinite bandwidth, coupled with massive computational power and AI-based human decision support, want to take that point to an extreme level. I believe this visionary concept has some merit, but in addition to not being very well-defined, or, as far as I can tell, adequately supported by analysis, the current concepts have at least two flaws: They don't ade-

---

16 The acronym has grown over time to BMC4 (adding Computers) and C4ISR or BMC4ISR or C5ISR (adding cyber). I'll distinguish the sensor, or intelligence, surveillance, and reconnaissance (ISR) suite, from the BMC3 network that connects elements of an operational concept and supports decisions.

quately consider the degree to which the enemy has a vote, and they overly value centralized human decision-making.

Americans generally talk about operational decision time in terms of the Observe, Orient, Decide, and Act (OODA) Loop. The ability to control time, to move and/or take effective action faster than the opponent, is derived from the collection of capabilities that affect decision speed and quality. It is a feature of C3BM systems' performance and the intelligence, surveillance, and reconnaissance (ISR) systems that feed information into those networks. In recent years, American military leaders have spoken *ad nauseam* about the need to "engage the enemy from a position of advantage at a time and place of our own choosing." Faster and better decision-making, in other words superior ISR and C3BM provides opportunities to achieve positional advantage and enables coordinated action—all to provide better relative operational efficiency. The point of superior C3BM and ISR is to enable greater lethality and survivability.

This is all fine, but my view is that in the US, our rhetoric on this has been a little over the top, going all the way back to digitization and network-centric warfare in the '90s, and now to multi-domain warfare. I agree that there is improved efficiency to be obtained by aggregating and analyzing information efficiently and providing better automated command and control tools, but the enemy has a vote. And the resiliency of this approach, as well as the limited potential efficiency improvements that are even theoretically possible, lead me to believe that the US may be over-relying on the concept of a military AI-driven internet of things (IoT) concept for future battlefield superiority.[17]

I have on occasion asked questions about the degree to which realistic or even ideal integrated C3BM can improve operational results. I know there are improvements to be had, but I'm concerned that the almost religious faith in this concept as a panacea is misplaced. How much more survivable will Joint All-Domain Command and Control (JADC2) make our aircraft carriers, forward air bases, and satellites? We need to quantify the operational impacts of JADC2 and more fully investigate the risks

---

17   I wrote this paragraph over five years ago; my views have not changed.

associated with responsive threats, especially kinetic threats to key nodes, as well as both cyber and electronic warfare threats. It's a sure bet that our adversaries are working on all these problems.

At this point, my greatest concern in this area is assuring survivable and effective ISR systems that can find and identify targets of interest with high probability of detection and low false-alarm rates—in all domains of warfare. Fusing sensor data can be very helpful at improving aggregate results, but doing so is technically challenging and requires secure high-data-rate communications and high-resolution sensors with reasonably good target detection and identification performance individually.[18] I think we can expect incremental improvements in this area, but merely applying advanced C3BM systems to existing operational concepts will not provide revolutionary military capabilities.

The other technology-driven area in which I think our rhetoric has been over the top, but less so, is AI. The basket of technologies we call artificial intelligence is truly transformative, and this transformation is happening quickly. I'm definitely not talking about true human-like AI, what we think of as consciousness.[19] I am talking about the forms of AI known as pattern recognition, machine learning, data analytics, autonomy, and executive decision-making or support. All these areas have been in development for decades, but the recent and continuing pace of advancements in these areas is what will make the concepts I will describe realistic and not science fiction. I believe many of them are possible in the relatively near term, less than one decade. As we field self-driving cars and trucks, deliver packages by UAV, and start to see fully automated (if not unmanned) commercial ocean shipping; as we see facial recognition at scale implemented on our laptops and through surveillance cameras;

---

18   One notable shortfall historically is the ability to find objects the enemy doesn't want found, especially stationary objects that can be disguised and for which cheap decoys can be fielded.

19   Since I wrote the original report, the introduction of large language models (LLMs) has resulted in much earlier predictions of true artificial general intelligence. We can increasingly program a computer to act like a sentient being, and we should definitely exercise caution as we approach that goal.

and highly sophisticated business systems surveilling and supporting the management of complex global supply chains, it is clear we are not far from the broad application of these technologies to military operations. We have the choice of applying AI technologies to improve the functionality and management of the systems and concepts we have today, or we can create wholly new systems and concepts based on those technologies. Our cultural bias takes us in the former direction. In the chapters that follow, I will attempt to go in the latter direction of new concepts and systems.[20]

AI technologies will be essential to all the concepts I will describe, but it won't be general intelligence; it will be in specific applications of some subset of AI technologies to specific military decision-making. Examples of those decisions include:

- How to best maneuver a specific system to survive and accomplish a mission.
- Reliably making an engagement decision within prescribed parameters.
- How to maneuver small numbers of systems together to best achieve an operational objective
- How to optimally structure and operationally deploy the building blocks of a concept to achieve a higher-level military objective
- How to best avoid collateral damage
- Where to place the boundary between an acceptable engagement and an unacceptable one
- The inference of enemy dispositions and intent from intelligence data; and comparative analysis of potential friendly and enemy courses of action leading to an optimized operations plan, air

---

20 I watched the Super Bowl while in the process of updating this chapter, which provoked the following thought. The book (and later, film) *Moneyball* introduced the idea that statistical analysis could significantly improve results for a professional baseball team. Let's go a step further and apply AI to play calling in football. This is basically a military-like tactical decision about what maneuver scheme to use against an opponent in a given tactical situation. I'm willing to bet that we aren't far from having algorithms that do this better than any coach, and in real time—if they don't exist already.

tasking order, or the equivalent in other domains and across multiple domains.

We are coming to the point where computers will be able to make these decisions, and many others, much more quickly and much more effectively than humans. When we get to the point where all the decisions that must be made on a platform will be better performed by AI, it's time to seriously consider getting the humans off the platform.

There is one more fundamental of warfare topic I want to mention before I turn to the central topic of this book. This fundamental will become less important as we progress because it can be dropped as all but irrelevant once one accepts the options that AI and lethal autonomy open up. That fundamental is human behavior and its limitations.

Let's start with courage—human acceptance of personal risk. This fundamental has been a constant in warfare since day one. It is embedded in our human fight-or-flight decision-making processes. Every military seeks to instill courage—a fighting spirit—the willingness to accept grave risk to self, the so-called warrior ethos, commitment to mission and to the unit, esprit de corps or "elan" as the French once emphasized.[21]

Most militaries like to think they have a superior ability to face danger and are better disciplined and more resolute than their opponents. In fact, these differences often, but not always, turn out to be ephemeral or insignificant. Human beings, especially taken in large groups, have a finite amount of courage available to spend. A lot of combat training and military indoctrination is about elevating this level of individual and group risk tolerance. Historically, units were considered combat-ineffective after receiving losses of around 10 percent. There are many examples of units being effective with higher losses, and some examples of military personnel willingly sacrificing their lives, not just individually but at scale, to

---

21 Secretary Hegseth has emphasized "restoring the warrior ethos". In my experience the US military has always had a strong warrior ethos, loosely defined as having the courage, skill and commitment to complete the mission no matter the obstacles or risks involved. This does not change with political administrations. It is deeply embedded in our professional warfighters, but all humans have inherent limitations.

attack an enemy (extremes include kamikaze pilots and terrorist suicide bombers), but this isn't the norm, and I don't think we will ever expect or demand this of Americans.

There are also plenty of examples of military organizations that lacked courage and fled when confronted by their enemy, but on the whole, it has not been a major discriminator in modern warfare. As much as militaries have tried to create an advantage in courage through training, selection, and leadership, there is not all that much difference among human beings. As a result, operational analysis has generally ignored this factor other than to note when units become theoretically ineffective. In theory, the willingness to move forward or to stay in place and engage the enemy could be a variable between zero and one. For a machine, it could always be one, or whatever conditionally based value is programmed into the machine. While courage is generally not considered a quantifiable variable in analysis of military modernization options, in practice, courage becomes an extremely important factor when comparing manned and unmanned systems and their operational abilities. Machines have infinite courage—again, if that is what is programmed into them, and this fact can be exploited operationally.

The other aspect of human behavior that is worth noting is the inherent limitations of the human body and mind. People must have sleep, food, and water to function, or their capabilities quickly atrophy. Prolonged exposure to trauma and stress will have an impact on how well people function. Experienced commanders are well aware of the importance of sleep discipline to unit performance. Machines need maintenance, but they do not need sleep. They also don't need food or water consumption, but in those cases, there are partially analogous needs for energy replenishment for machines. Machines also are not traumatized by exposure to prolonged combat. Finally, human life support places limitations on design parameters like g-forces, temperature, atmospheric environments, strength, and reaction time. If we can envision weapon systems without all these constraints and with military decision-making and behavior superior to that of humans, then we should certainly explore that possibility.

This brings us finally to autonomy and the topic the Pentagon currently tries to avoid if at all possible—lethal autonomy. Autonomous lethal systems will be a central focus of this book. Let's discuss the acceptability of taking this approach.

LETHAL AUTONOMY

Importantly, both modern American "offset" strategies essentially utilized existing force structures and platforms and simply added new munitions; in one case tactical nuclear weapons and in the other precision munitions. From the point of view of cultural operational change, both fit well into existing cultural norms and didn't require fundamentally new roles for operators. The US just used the same or very similar manned platforms to deliver the new munitions to the enemy. Today the emphasis for modernization is on multi-domain operations and JADC2 with ubiquitous high-bandwidth connectivity and AI-assisted battle command. JADC2 and its military service counterparts also have the feature of overlaying well onto current operational concepts and weapon systems. This isn't a coincidence. I think we can go much further in the next decade, but more fundamental changes will be required to do so.

Most recently there has been an increased emphasis on uncrewed systems, especially small drones. To date, Bob Work's vision of the autonomy-based Third Offset Strategy has not been fully embraced, although experimentation is proceeding across DOD and there is a rush to emulate the drone proliferation we have seen in Ukraine. I believe this is the right general direction, but we need to move more thoughtfully and more decisively. As Secretary of the Air Force, I did everything I could to move this idea forward. Most notably, I initiated a program called Collaborative Combat Aircraft, or CCAs. While all of the military services have been experimenting with autonomous systems for years, the CCA program represents the first major commitment to lethal autonomy. Earlier concepts, such as Reaper uncrewed aircraft, had been remotely piloted and all engagements were under direct human control. The CCAs will still be under human supervision, but their potential degree of autonomy will be unprecedented, and they will be an integral part of the force structure.

The military services are all experimenting with uncrewed systems in various roles and fielding experimental prototypes on initial complementary platforms. None, however, have embraced or even seriously considered autonomous lethal unmanned systems as an alternative to manned systems. Even the CCA program that I initiated envisioned a crewed fighter controlling several CCAs rather than CCAs operating completely independently. Prior to the CCAs, the closest the Defense Department has come to fielding lethal autonomy would seem to be the use of unmanned air vehicles such as Reaper for reconnaissance and intelligence with very tightly human controlled lethal engagements against terrorists.[22] It can be argued that missiles with various post-launch autonomous target acquisition capabilities cross this line, but the DOD has not considered these systems to fit within the lethal autonomy definition. The US has been very reluctant to explore fully autonomous lethal unmanned concepts, largely because of a strong reticence to accept the consequences and implications of lethal autonomy. There are several dimensions to lethal autonomy, and it's worthwhile to explore them before positing new operational concepts built around the acceptance of at least some degree of lethal autonomy. Let's start with some of the barriers to the use of lethal autonomy and then turn to the potential advantages lethal unmanned systems can provide in terms of the variables discussed above.

When most people think about lethal autonomy, if they think about it at all, they probably have in mind some robotic device with human-like artificial intelligence that may "go rogue" and kill indiscriminately or

---

22 Deputy Secretary Kathleen Hicks announced the Replicator program in
2023 with the goal of fielding large numbers of lethal autonomous systems
in all domains, but very little information about the content of the program
or the degree of lethal autonomy included has been made public as of
this publication. "DOD Replicator Initiative: Background and Issues for
Congress," Congressional Research Services, August 25, 2025,https://www.
congress.gov/crs_external_products/IF/PDF/IF12611/IF12611.9.pdf.

recklessly. This "Terminator" concept begets an emotional reaction and has motivated a lot of smart prominent people to recommend some kind of an international treaty banning nation-states from fielding lethal autonomous systems.[23] In my view, this perspective and this recommendation are inaccurate and impractical.[24] Lethal autonomy in some form has actually been with us for a long time. Lethal autonomy isn't and doesn't need to be human-like in any meaningful way. Here are some examples.

The Improved HAWK medium-range air defense system I operated in the 1970s had an "automatic" mode that would have enabled the system, using a truly primitive computer by today's standards, to assess possible threats, apply rules of engagement, and fire missiles at threats deemed hostile, all without human intervention. The idea, even in the '70s, was that in a dense air attack situation, human operators could not keep up with the decision-making speed required to prioritize and engage individual threats. Almost fifty years later, our current air engagement systems are infinitely more sophisticated and capable than the HAWK system I operated in the '70s. Threats are also much more sophisticated in their capacity to work together and employ countermeasures. Human beings still have the same limitations.

A few years ago, I had the opportunity during a trip to Israel to participate in a demonstration of the self-defense capability on a Merkava tank, an active protection system called "Trophy." Once enabled by an operator, the system is highly autonomous. It detects a threat launched at the tank, tracks the threat, and commands an engagement with a kinetic-kill interceptor that is launched from a container on the tank. All of this has to happen so quickly that human intervention isn't practical. Because this

---

23 Guia Marie Del Prado, "Stephen Hawking, Elon Musk, Steve Wozniak, and over 1,000 AI Researchers Co-Signed an Open Letter to Ban Killer Robots," *Business Insider*, July 27, 2015,https://www.businessinsider.com/ stephen-hawking-elon-musk-sign-open-letter-to-ban-killer-robots-2015-7.

24 The National Security Commission on Artificial Intelligence report argued that the US had to develop lethal autonomy because our adversaries would not be constrained from doing so. I agree with this assessment. https:// cybercemetery.unt.edu/nscai/20211005220330/http://nscai.gov/.

system is defensive, the use of autonomy and the launch of a munition is accepted, although precautions must be taken to protect any infantry or civilians near the tank. There is one other notable feature of this system, however. When Trophy detects a threat, in addition to conducting the intercept of the incoming missile, it also orders the main gun and coaxial machine gun to the point of origin of the threat. There is one more step that would be needed to automate a lethal response—pulling the trigger. Technically this is a trivial addition to the autonomous capability already in place. There is no real barrier to implementing an autonomous lethal capability, on this or on other weapon systems.

Anti-ship missiles with active and passive seekers that provide for the guidance of the missile to its target have been with us for a long time. Over decades these missiles have become more and more sophisticated. As the range from the launch platform to the target has increased, the need for autonomy has also increased. Ships move, and if one is trying to attack the higher-value targets (e.g., aircraft carriers) in a formation of ships with mixed military vessels, commercial shipping, and perhaps decoys, then one needs a smart seeker that can scan the relevant ocean surface and pick out the right target, and even the optimal aim point on that target.

Taking this one step further, if the attacker is trying to penetrate layers of soft- and hard-kill defenses to get to the high-priority targets, it makes sense to share data among the attacking missiles and to automate cooperation to maximize their joint lethal effects. This isn't just lethal autonomy, it is collective lethal autonomy, and to a large degree it already exists.

Given the reality described above, arms control limitations are problematic. As a humanitarian and a human rights attorney, I'm all for effective arms control, but effective means meaningful and enforceable. Given that lethal autonomy is essentially with us now, as the examples above show, I don't see how an effective arms control regime is possible. One has to begin with definitions of the things one wants to ban or constrain. I don't know how to draw a line between the supposedly acceptable types of systems I described above and those that would be banned. Trying to distinguish defensive (good) from offensive (bad) weapons doesn't work when an autonomous air defense system can engage a civilian airliner. Most weapons can be used both defensively and offensively; killing is killing. Trying to distinguish degrees of autonomy is problematic as there is no obvious place to draw a well-defined line between permissible and impermissible technologies or capabilities. The problem here is that there is a great appetite for technologies that automate functions like target detection, enemy intent inference, and optimum engagement planning. These capabilities are continuums with no sharp boundaries. There is a sharp boundary between preparing to engage and actually pulling the trigger, but, as with the Israeli Trophy system I observed, that boundary is technically trivial and crossing it would be easy to conceal. One possibility might be to ban all unmanned systems from carrying weapons. This approach would ban existing lethal drones, which one could consider, but it would not solve the problem of the risks associated with lethal autonomy. A system which includes a human operator can still operate in an autonomous mode, like my Improved HAWK air defense system from the '70s. It would also tend to ban existing cruise missiles and other fire-and-forget munitions systems with autonomous seeker functionality, something which I expect no major power would accept. I'd be very willing to engage in a conversation aimed at solving this problem, but I'm not optimistic.

Assuming one could draw a line, there is still the problem of enforcement. The difference between full lethal autonomy and human control is some mechanism by which a human decision must be made to permit an engagement. Think of this as a trigger. In modern electronics systems, a trigger is a switch, which is more likely to be digital than mechanical. It is also likely to be remote from the lethal mechanism itself and the decision communicated through a digital link of some kind. I haven't heard anyone suggest that it isn't acceptable to automate everything but the crucial engagement decision. As a result, "cheating" on a ban on lethal autonomy is simplicity itself; somewhere in the weapon system include the capacity to close that switch autonomously. The only way to prevent this from being done covertly is to have extremely detailed design information and extremely intrusive inspection regimes. Even if nations were willing to permit this, which they are not, it would be next to impossible to implement.

I would also question whether restricting the availability of lethal autonomy through international law is even wise, assuming it could be done effectively. Is there some good reason why we'd prefer that humans kill and be killed in war as opposed to machines assuming these functions? Over time many unpleasant functions humans performed have been turned over to machines. This includes the mind-numbingly repetitive and those involving brute force. A lot of human progress has been about relieving humans of functions that can be done less painfully and more efficiently by machines—freeing humans to engage in other pursuits. It may sound fanciful, but if we have to resolve our conflicts through competitive violence, why not let our machines do that for us? This might strike some as ridiculous, but what is really ridiculous to me is that we still have a preference for sending humans out to kill each other to resolve our differences.

I believe there is a better and more pragmatic way to constrain human rights abuses than to ban lethal autonomy; it is to hold the responsible people accountable, regardless of the tools they employ. We have reasonably well-defined laws governing use of force now. The law of armed conflict may not be enforced consistently, but it does provide established and agreed limits to the brutality of war and the application of violence. There is no reason why the people who unleash indiscriminate autonomous killers on innocent victims, including neutrals and noncombatants, cannot be held accountable for their behavior. At the end of the day, we want to constrain the behaviors that we find objectionable. I believe we can still do that without banning autonomous weapons.

It would certainly take some effort by legal theoreticians and others to determine who in the chain of control of an autonomous lethal system to hold accountable and what the standards for violations should be for specific acts. I would offer that a good place to start is by enhancing the standards we apply today. Currently, we accept that some "collateral damage" is inevitable in war. We ban collateral damage that is not a military necessity, intended to inflict terror on civilian populations, or intended to achieve genocide or ethnic cleansing. If we are going to permit autonomous machines to make decisions about lethal engagement, shouldn't we raise the bar for when that's acceptable? In any event, the minimum is the standard we employ today, and there is no reason to relax it.

So, who should be held responsible? This is a matter that is very standard in criminal law and personal injury cases; determining intent or degree of negligence required for liability. Any person or country that uses an autonomous lethal system should have a legal duty to ensure that system will not violate the law of armed conflict. This means that the designers, the testers, the acquirers, the fielders, the military operators, and the political decision-makers, all have a duty to ensure that a fielded system is compliant with established rules governing operational behaviors. In some cases, strict liability could be implemented. It might be useful to also require that a specific designated official and organization be responsible for testing and verifying compliance, analogous perhaps to air worthiness certification today. For egregious violations—the intentional wholesale slaughter of innocents—the humans who directed or even permitted this use would be responsible, just as they are today.

As a last point on this topic, I don't think the development of lethal autonomy can be stopped or even slowed appreciably. Most of the enabling technologies will proceed for independent reasons. The military advantages of employing lethal autonomous systems will become increasingly apparent to all. As I look at just ongoing US military service and DARPA efforts in this area, it is clear advances that will take us further into the world of lethal autonomy are well underway and advancing rapidly. Some nation-states and groups will not hesitate to embrace this opportunity, regardless of any ethical concerns. I've heard representatives of several liberal democracies state publicly they will pursue the relevant technologies because of the fear less principled nations will do so. Almost daily I read open literature accounts about new unmanned lethal systems or improvements in the relevant areas of artificial intelligence—especially pattern (target) recognition. This genie will not be put back in the bottle, so let's look into the bottle and see what might be inside.

If one can get past the objections to lethal autonomy, it opens up a fascinating range of options, including tactical behaviors that would certainly not be doctrinal for manned systems. Specifically, it is generally permissible, if the cost is not prohibitive, to sacrifice unmanned systems, singly or in groups, for a tactical or operational advantage. My late friend, retired Army Major General Bob Scales, liked to talk about how a fundamental function of infantrymen, particularly in counterinsurgency campaigns, is to walk down a trail with a rifle toward the enemy in order to draw fire and expose the enemy to attack. That's an oversimplification, but there is some truth in it. I once convened a group of general officers in the Pentagon to try exploring ways to end this paradigm. We didn't get very far, but the option of using relatively inexpensive and expendable unmanned systems for this purpose is very attractive to me. The Ukraine conflict has shown us that it isn't just attractive, it's required. In the chapters that follow, I will apply this intent to each of the military domains: land, sea surface, subsurface, air, space, and cyberspace.

COMMON CONCEPTUAL THREADS: THE RELATIONSHIP BETWEEN PROJECTILES, LAUNCHERS, AND TRANSPORTERS AND A KEY ROLE FOR HUMANS

I started to write this section with the hope that across domains there was a common way to describe weapon systems and operational concept components, and a common way to talk about an optimization paradigm. My thought was that there might be a common solution, conceptually at least, in the trade between projectile range and cost, the number of projectiles in a launcher and its cost, and the capacity and cost of a launcher's transporter (where one was required). I was thinking, for example, of the arsenal plane or arsenal ship concepts and of an unmanned ground vehicle (UGV) carrier concept. All would be attractive solutions demonstrating that unmanned lethal autonomy warfare operational concepts in one domain would have similarities to other domains. After some effort,

I couldn't convince myself this was clearly the case, but it still seemed a useful way to discuss concepts in most domains.

Also, while there is commonality in the implications of some key parameters, each domain brings its own unique characteristics. In particular, it is easier to survive in some domains than others, so one can venture closer to the enemy with larger, more expensive platforms. It is also easier to communicate in some domains, affecting the ability to use offboard sensors, fuse information, and coordinate platforms and weapons. While the tentative domain-specific concepts I will suggest may not be similar, the concept of projectiles, launchers, and transporters has validity, and most domains can be approached from that perspective. Alas, warfare is complicated, and there is no avoiding the complexity of the specific design trades involved, but there is at least a common way to approach or think about the problem.

As I worked my way through thinking about and writing about each domain and multi-domain operations, one common concept that did emerge was in the role of humans. In every case, I found a need to include humans at an organizational level somewhere above individual engagement decisions and small-unit tactical behaviors, but generally not far above those levels (space and sea subsurface are potential exceptions). Humans would be needed to control operations at that level; they provide an executive decision-making function for force management. At this level, time constraints are less severe, making processes at human information-processing and decision-making speeds operationally acceptable. I do envision robust AI-based decision support tools at this level, but not complete autonomy. Against responsive threats and in a broad range of operational contexts, something closer to the general intelligence humans have is needed to at least oversee operational decision-making. AI will not provide that anytime soon and I doubt that humans will be willing to cede that level of control for some time to come.

In the balance of this book, I will be discussing various warfare domains. In all domains, warfare is largely about applying destructive force to something one wants to destroy without being destroyed in the process. Simply and abstractly, energy must be delivered by a projectile to a target to kill that target. A launcher provides the deployment mechanism for the projectiles. Another object, a transporter, may be necessary to move launchers to where they can be employed. Importantly, both launchers and transporters are generally reusable, whether they are crewed by humans or not. Projectiles are generally expendable and consumed when used.[25] This can be a useful construct because so many of the design trades involved in a concept are about the balance of fundamental features among these elements. The range of the projectile and the payload it carries (warhead and guidance, for example) drive its mass and other characteristics. The projectile mass and dimensions drive the launcher's payload capacity, as does the stowed inventory and the range of the launcher. If a transporter is involved (aircraft carrier, for example), that platform in turn is constrained by the mass and dimensions of the launchers and stowed projectiles the transporter must bring within range of the projectiles and launchers. Obviously, the laws of physics constrain everything, as do several other factors, and sometimes there are more, or fewer, nested entities involved than two or three. Everything in a concept of operations can't be shoehorned into this model, but it's a useful place to start when, as here, one is working with a clean sheet of paper.

The fundamental objective is to apply energy to an enemy target and destroy or neutralize that target. That energy is delivered by some form of a projectile, which can be a bullet, a missile, a torpedo, or the photons in a laser beam. Projectiles are consumed in use, so they need to be as inexpensive as possible. We'd also like them to be long-range, resilient, and have high kill probabilities, all of which adds cost. Projectiles are never manned, even today.[26] These projectiles are sent on their way to the target by some sort of device that we can call a launcher. I'm thinking of a

---

25   Reuse is a key component for cost-effectiveness. As an example,
SpaceX revolutionized space launch economics when it introduced
reusable rockets. For reuse to have operational value, it must
be accompanied by a reasonable level of survivability.

26   There are exceptions, for example, kamikaze-piloted
aircraft and suicide bombers as noted earlier.

launcher as a platform, not just the launching device on a platform. Today, this can be a tank with a main gun, a tactical aircraft with internal or wing-mounted launch points, or a surface ship from which the projectile—the bullet or missile—originates. Launchers may or may not be reusable, and they may or may not be manned. In my concepts they generally will not be manned. If launchers are not reusable, we want them to be inexpensive also. If reusable, they need to be survivable through some combination of standoff range and other features, which again adds cost. Since we are assuming unmanned launch platforms, we can also talk about attrition being more acceptable, so there is a more open design trade space than we would consider for manned systems.

Finally, the launcher has to be delivered to a place where it can be utilized so that the associated projectiles can be effective against the targets of interest. We can call that device a transporter. Some launchers double as transporters—they self-deploy, at least to some degree. Others must be moved to within ranges where they can operate tactically. This can be done through prepositioning, such as on forward air bases, in pre-positioned depots, on bases close to the expected area of operations, with locations dictated by the operational range of the launchers, or by transporters like aircraft carriers, surface ships, trains, and trucks. I will make some judgments, valid or not, about the threats to and the survivability and resilience of the transporters, but like it or not, they are necessary and must be considered part of the operational concept. Because the United States' conventional forces are intended for power-projection applications thousands of miles from our shores, we have a particular requirement to consider how we will get to, or already be, where the fight will occur.

This hierarchical system can be thought of as a nested set of systems. An aircraft carrier is a transporter. It brings launcher airplanes to where they can reach close enough to the enemy targets that the projectiles they carry can reach the targets of interest. Historically, we rarely, if ever, design all the elements of this suite of systems at once so that they can be optimized relative to one another. We are almost always trapped by the existence of large prior capital investments that constrain us. The new missile must be compatible with the existing aircraft; the new aircraft must be compatible with the existing aircraft carrier and the existing missiles—and so on. We should take advantage of the opportunity autonomy and other emerging technologies give us to rethink the entire design of our future warfighting concepts. We ultimately may not completely break free of our past investments, but we should at least start our thinking and analysis unconstrained by past decisions.

One of the limiting factors in the existing nested projectile/launcher/transporter designs has also been the need to include human beings on the transporters and launchers, for a wide array of functions. A tank is a good example. If the functions of driver, gunner, loader, and commander must be performed by humans, the tank design must incorporate protected volume for these people together with the tools they need to do their functions and the supplies they need as human beings. In addition, the launcher must accommodate weapons (main gun and auxiliary weaponry) and a reasonable number of projectiles. As we automate the functions of the crew, we can eliminate some of these constraints incrementally. If we eliminate the crew entirely, we can reimagine the whole trade space as well as the range of possible tactics and operational techniques. In every domain, once we introduce uncrewed systems, the projectile, launcher, and transporter trade space can move to a new optimal design point. We can at least approach the problem in each domain by thinking about the trades between these functions.

We can start this process with a known entity. The amount of energy it takes to defeat a given target is well understood. The projectile's job is to get that energy to the target. That energy is the minimum payload our projectile has to carry. The other feature that sizes the projectile (think missile) is its range. The farther we want the projectile to travel, the bigger and more expensive it will be. Because projectiles are generally not reusable, once they are launched, they are expended.[27] Again, the things we

---

27 Conceivably, projectiles that can't find acceptable targets could return, be refueled, and be reused, but the cost of including this in a projectile design is likely only justified if a large fraction of the projectiles will not find and attack targets.

will consume instead of reuse should be as cheap as possible. Therefore, we would like the launcher and carriers to be able to get close to the targets so that we can use short-range weapons and carry more of them per reusable carrier. Unfortunately, it's not quite this simple; projectile range also adds targeting flexibility, which has value, and improves the survivability of the launcher, so there is some appetite for projectile range. We also want as many projectiles as possible to arrive at their targets successfully, so there is a cost-to-complexity ratio that has to be optimized in a trade that balances the numbers of projectiles needed and the probability any given projectile gets to the target to minimize the total cost per kill. Hypersonic weapons are a good example of trading high cost for high probability of successful penetration to the target and, hopefully, successful target engagement. Moving up the chain from the projectile to the launcher to the transporter, we can start defining concepts that have some relevance and optimal balance of cost and functionality for each domain.

Many current ideas for autonomous unmanned weapon systems fail to consider the need to get those weapons (launchers and projectiles) to where they can be useful. How are they transported to the operational area in a survivable, reliable, timely, and cost-effective way? If pre-deployed, are our assets survivable against surprise or warned attack? If not, how will we get them where they need to be? If we are assuming high rates of consumption or attrition for projectiles and launchers, how will our losses be replaced over even a moderately protracted conflict? The trade space for weapon systems design, and for operational concept design, involves all three of these functions, with many possible variations and combinations other than the simple three-part harmony just described (projectile, launcher, and transporter). This trade space manifests itself in decisions about how many projectiles on what types of launchers and what types and capacities of associated transporters (if any) are needed to complete and to optimize overall cost-effectiveness. We can certainly combine these functions, and we don't need to be limited to only three components.

In the concepts for each domain, we also have to provide for other necessary functions besides delivering lethal energy to targets of interest. These include surveillance and targeting, battle management, communications, command and control, and especially sustainment. I will touch on these topics and other domain-specific complexities in the sections that follow.

Sustainment, in particular, is a major cost driver and as noted above often neglected when operational concepts are considered. Relying on autonomy opens new options and creates challenges for sustainment. Consider either sustainment model (American or Soviet) discussed earlier as compared to one that would be possible if most of our fighting platforms were unmanned. We would need to use some fraction of these platforms in peacetime for experimentation and to train the AI tactical software that controlled them, but this could all be accomplished with a small number of platforms; most of our assets could be idle and not undergoing the wear and tear of frequent or continuous training. Upgrades could be done much more efficiently. Software changes (most functionality will be software-defined in the future) could be done across the force very quickly. Hardware upgrades, which would be substantially less frequent, could also be scheduled much more efficiently because we wouldn't need to schedule around human training cycles and deployments. As indicated above, at some level of the force structure, there would still be operators whose role would be to manage, oversee, and control the autonomous platforms and the small units composed of autonomous platforms, but this would be done remotely from a small number of platforms or command and control nodes. Realistic simulations, anchored by constant experimentation based on field experience data, would provide a fully adequate low-cost training environment for these operational executive-level people. The benefits to peacetime cost and reduction in support structure would be enormous. There is a valid engineering trade space between sustainment related design features, cost, and operational effectiveness that would be opened by reliance on unmanned autonomous systems. The optimum design choices would consider wartime expected life and design for high reliability after long periods of storage and inactivity. So called "wooden round" munitions have been designed for highly reliable shelf life for some time, but less stringent requirements would probably be adequate, even optimal, for more expendable or attritable unmanned fighting platforms (launchers and transporters), especially if most faults were detectable autonomously with high confidence at system start-up. Life-cycle costs and both peacetime and wartime sustainment requirements have to be part of the trade space.

RESPONSIVE THREATS AND ENDURING ADVANTAGES

Every innovative military development provokes a response. Introduction of the new-model forces I'll describe in the following sections would cause a strong and immediate reaction. There are three relatively short term reactions to innovation on the battlefield: rapid reaction, deliberate countermeasures, and emulation. A good relatively recent example of the US responding to a new threat was the introduction of improvised lethal small UAVs at scale by ISIS against the Iraqi Army and US special operators in 2015 and 2016. The first phase of a reaction would be similar to that of the US, a rapid acquisition program and responsive tactics to address the new threat. For the ISIS unmanned aircraft system (UAS) problem, this took the form of the rapid introduction of urgently procured soft- and hard-kill small-UAV defeat mechanisms into the force, even if limited in their effectiveness and overly expensive. This is followed by a more thoughtful and sophisticated approach which may include novel and new designs. In the counter-UAS case, examples of this included research on special-purpose smart proximity rounds, purpose-built target acquisition and fire control systems for existing guns, high-energy lasers, and area weapons employing high-power microwave engagements. These responses, in turn, lead to more sophisticated small-UAV threats—with some combination of hardening, deception, and counter-kill.[28] More significantly, they lead to emulation—the adoption and extension by others of the new technology and operational concepts, such as the widespread adoption of small armed UASs, which we are seeing today in Ukraine at a massive scale and which I discuss as part of the concepts I will describe in this book. This, in turn, leads to a period of incremental improvements on all sides until, eventually, a new revolutionary concept is enabled by emerging technology and operational innovation.

As technology changes warfare, advantages can have varying degrees of longevity. Some changes will be emulated immediately, with lead times measured in months or at most a few years. To the extent the technologies involved are militarily unique and difficult to copy, advantages can be of longer duration. The recent example of the US advantage in precision munitions, stealth, and wide-area surveillance had a relatively long life, arguably about two decades, before the same technologies were adopted successfully to be used against the US. Stealth has been particularly enduring. What these technologies have in common is that they are military technologies with little or no commercial application. The greater the depth of knowledge required and the more specialized the technology, the longer an advantage based on it lasts. It's comparatively easy to control access to the militarily unique technologies because they can be kept concealed during development and are harder to emulate or reverse engineer once revealed. Even for these technologies, the most important piece of information—that they exist—cannot be concealed for long. It's easy to focus research and espionage efforts once an existence proof has been provided by an adversary. Consider, for example, how long the US nuclear monopoly lasted. The technologies that form the basis for this book, on the other hand, such as those associated with autonomy, some of

---

28   This is a good example of why commercial products don't have long operational lives. Once the adversary starts finding ways to attack the commercial products—for example, through jamming, cyber, and directed energy—the militarily unique requirements start to pile up, and before long, the commercial product is replaced by something much more highly specialized.

the artificial intelligence technologies, and modern information systems technologies, are all commercially based and will be widely available. Any advantages obtained from them will be relatively short-lived and will lead to more militarized versions of those technologies very quickly. The game goes on, and as long as humans choose to resolve their differences by violence, it will never end.

DOMAINS OF WARFARE AND MULTI-DOMAIN WARFARE

At last, we come to the heart of the matter: what future warfighting operational concepts will consist of, and what systems will be included in them. There are no clean domain boundaries, and one can argue endlessly about definitions. The intent is to be pragmatic, not tautologically pure. In addition to land and air domains, I've divided sea into surface and subsurface. I've also added space and cyber as domains. I did not add the electromagnetic spectrum or electronic warfare (EW) as a separate domain but chose to handle that in the context of the other domains.[29] Cyber is discussed both in the context of each domain and as a domain in its own right. My logic was to think of domains as abstract "terrain" on which forces are organized and equipped at scale to contest for control of that terrain. In each case, the physics of the terrain has features that are major drivers for the forces involved. It's easy to quibble about these choices. I was looking for a reasonable organizing structure, not a definitively correct answer. Readers are asked to accept, if only for the purpose of this book, that the chosen list of domains is a reasonable basis upon which to lay out some concepts and ideas for the future of warfare. I will discuss multi-domain implications in each specific domain chapter, and I have included a chap-

---

29   There is an ongoing debate about this that I don't have strong feelings about either way. Control of the electromagnetic spectrum is hugely important to warfare in virtually all domains, undersea being up to a point an exception. What's important is understanding EW's significance and potential decisiveness in any domain and that it cannot be separated from considerations about how to prevail in each domain.

ter dedicated to multi-domain, joint, and combined operations at the end of the discussion of the individual domains.

For each domain, I'll provide a short introduction and discuss the operational concept I envision and the building-block components of the concept. I'll then cover a list of complexities associated with each domain. The lists tend to be long; warfare is complicated. I've organized the so-called complexities somewhat arbitrarily by three categories: (1) fundamental operational needs, (2) design requirements and considerations, and (3) other needed military functions. Within each category, topics are arranged alphabetically. For each domain, I include cross-domain considerations under fundamental operational needs and responsive threats under design requirements and considerations. We should expect and even demand significant cross-domain functionality. All domains will have cross-domain complexities; they will take the form of cross-domain dependencies, support, and effects. There is also great virtue in military concepts in having redundant ways to do anything; redundancy provides flexibility, enables surprise, eliminates single points of failure, improves resiliency in general, and compounds the enemy's problem dramatically. Each domain will also have complexities unique to that domain.

Throughout all this, the design trades are complex and have to be done in detail in each domain to reach anything like optimal results. All I can do here is lay out some intuitively interesting possibilities, anchored in a lifetime of work on the intersection of technology, military systems, and innovation. If we discard our preconceived notions about what platforms and weapons should look like, and embrace the potential for widespread use of autonomy, we will enter a brave new world of conceptual design. I can only begin to explore that world here, but my hope is that this exercise will help others to think constructively about the possibilities and lay the groundwork for more quantitative analysis. Let's get started.

CHAPTER 2

WARFARE DOMAINS– LAND

Land warfare is about taking and holding ground. It is the oldest form of human warfare and has continuously evolved. That evolution has been driven by changes in weapon range and lethality, degrees of protection against those weapons, and mobility. Mechanisms for acquiring information, command and control, and logistics related technology have played important roles as well. The ground environment provides natural protection, both cover and concealment, restricts movement, and limits lines of observation and fire. Urban environments are especially close and complex, leading to high casualties. Ground combat has always been intensely dense with human participation and direct action between human combatants, and casualties at the point of contact tend to be very high. In the not too distant future, while the environment and its implications will remain, the direct participation of humans will decrease dramatically.[1]

---

1   A lot has happened in the five years since I originally wrote this paragraph, but my views have not been fundamentally altered by the events in Ukraine or elsewhere. There has been significant progress in the use of uncrewed platforms, air and ground, in support of ground operations, but it has been driven by the urgency of the moment and the current availability of off-the-shelf technology. For an excellent discussion of why Ukraine does not provide the last word or an enduring model for

In the following sections, I will first describe the overall concept to give readers a general understanding of the intent. I will then go into more detail about the key building blocks of the concept and then cover a range of complexities that must be addressed in designing an actual and complete military force. Ground combat is extremely complex, so this last section will cover a lot of ground.

OPERATIONAL CONCEPT

As in other domains to come, we will start with a blank page. There are no squads of infantry soldiers or platoons of manned armored vehicles. Those concepts are going to be untenable. Humans, as massed infantry or in tanks and armored personnel carriers are too valuable, too expensive, too targetable, too slow, too vulnerable, and not capable enough to survive in combat against the types of forces and formations envisioned here. Because of its relative complexity, the land combat domain is the hardest domain in which to make lethal autonomous forces practical, but it is going to happen. Fundamentally, we need three technology-based capabilities, all of which are more difficult in this domain than others: autonomous mobility, autonomous small-unit tactical behaviors, and high-confidence target acquisition. I've watched all these technologies mature over many years, and I believe they are close enough in reach now that this is a realistically foreseeable capability. What I will describe is an operational concept, a new model force that completely dominates current conventional ground forces. We are seeing the beginning of the transition to this type of force in Ukraine, but only the beginning. The force envisioned can conduct either offensive or defensive operations, but it is especially effective at defeating maneuvering conventional forces, mechanized or infantry. I will describe how this force operates to defeat a conventional modern human-centric force. Let's jump right in.

---

future ground forces, see: Vitaliy Goncharuk, "Ukraine Isn't the Model for Winning the Innovation War," *War on the Rocks*, August 12, 2025, https://warontherocks. com/2025/08/ukraine-isnt-the-model-for-winning-the-innovation-war/.

The basic concept is an echeloned force that combines ground and airborne unmanned autonomous systems operating in conjunction and in teams, with a nested command/management structure that is very adaptable, tailorable, and dynamic. At the lowest levels, where tactical engagements occur, decision-making is highly automated based on rules of engagement and temporal and spatial constraints. As one moves up to higher levels of force size and integration, human decision-making is introduced as it becomes more necessary and effective. The volume of information flowing from lower to higher echelons is minimized and tailored to support the decisions best made at a higher level of aggregation. In any many-on-many dense scenarios, this would not include the highly detailed information to support individual engagement decisions by humans, or even to support small-unit tactical decisions by humans. It would initially include information to support force management decisions and operational scheme-of-maneuver decisions, but even lower echelon (think small unit, company and below, at least) decisions in these areas could be largely automated. This structure can also learn autonomously at each echelon level and adjust both tactics and organizational structures in near real time in response to threat situational awareness and observed threat behaviors. This machine learning capacity is already being enabled by the type of technology and AI found in large language models that have been developed in just the last few years.

Within the new ground force model, medium-sized armed airborne UASs (launchers) with vertical takeoff and landing capability are launched from a ground UGV mother vehicle (tactical transporter) that contains multiple, possibly stacked, vehicles that provide lethal effects, target detection, and identification. The UAS participates in cooperative distributed tactical decision-making at the lowest echelon level, operating collaboratively in small units (nominally, say, four air vehicles launched from the same ground vehicle). The lowest echelon size is flexible, but it could be anything from two air vehicles to, say, ten. The air systems provide airborne line-of-sight engagement capability. While reusable armed medium-sized UASs are the workhorse of the concept, smaller UASs could also be included, especially on UGSs operating in closer proximity to enemy forces. Other autonomous ground systems provide primarily indirect-fire capability and ground transportation, energy, and munitions reload support to the unmanned aerial systems. Multiple ground vehicle configurations would probably be necessary to keep the individual ground vehicle sizes optimal. Different configurations could include indirect fires (launcher/transporter role) and air vehicle transport and support. One configuration could be a direct-fire capability, and all UGVs might have a minimum self-defense capability to provide resiliency. I'd categorize both air and ground vehicles as attritable but not expendable. Human-crewed command and control vehicles would be part of the mix and would be designed to be more survivable, mostly through stealth, and employ tactics to avoid being engaged.

Above the first echelon level would be higher aggregations of organization and automated behaviors, using the lowest echelon as behavioral organizational building blocks. The analogy to squads as the lowest level followed by platoons and companies, all with similar organic capabilities, is reasonably valid as a starting point to visualize this, but the model force does not have this rigidity. Tactical behavior models that would be automated would extend two, three, or more echelons above the lowest echelon. The force model includes an automated or semi-automated hierarchical C3BM system at each echelon that directs both the organizational structure and the employment of the echeloned forces assigned to operate under its control as a unit. Again, the analogy to current echeloned operational planning is apt, except that, in this case, it is much more flexible, automated, and tactically optimized to suit the immediate operational situation. The US Army has agonized for decades over the optimal size and composition of an infantry squad, platoon, and company, and how to best train the humans in those units. In this new largely uncrewed model, training requirements for humans are not a limiting factor for organizational design.

The intent of this structure is to overwhelm and destroy an opposing mechanized or infantry force in any given assigned area with highly favorable cost-exchange ratios. Swarming does not adequately capture the intent. The idea here is more sophisticated and threat responsive, although there would be some similarities. At the lowest tactical level, small numbers (nominally five to ten) of airborne systems would scour an area of interest and, depending on the results, choose an operational mode most efficient for destroying an enemy force or targets in the area. This function would be used in an offensive, defensive, or movement-to-contact mode as dictated by the mission. Specifics of tactical behaviors and resource management would vary depending on the situation, however.

The preferred engagement in this concept is by indirect fires or UAV direct fire (which could be just beyond line of sight). I have made some provision for direct-fire capability on ground vehicles above for resiliency, but the basic idea is that individual or ground vehicle-based line-of-sight fires have gone the way of hand-to-hand combat. In this concept, those engagements are not quite obsolete, but they are not preferred, rarely occur, and are avoided if possible. We have seen early versions of this reliance on drones and indirect fire in both Nagorno-Karabakh and Ukraine. At this point, they haven't completely displaced traditional mechanized forces, but the trends are clear.[2]

For a conventional opposing human-centric infantry and armor force of any type, direct-fire engagements from the medium UAS would be one option, as would be coordinated indirect fires from ground platforms within a few kilometers but beyond line of sight of the opposing force. The air vehicles in the concept would support indirect fires with targeting-quality information. Target identification would normally be an organic capability of the air vehicles within established rules of engagement and confidence levels, minimizing time delays and communication loads. In some lower-intensity situations, dual redundant target ID and/or other safeguards could be mandated to reduce false positives and engagement of civilians or wasted munitions. The force would seek to optimize the mix of weapons and the exposure of air vehicles to enemy lethal mechanisms to

---

2   For an excellent discussion of the use of drones in Nagorno-
     Karabakh, see: Antal, *7 Seconds to Die.*

be most cost-effective at each echelon. The immediate operational depth achievable at the lowest echelon by the organic air vehicles would be a few kilometers. The endurance of the air vehicles in the concept would be consistent with operations at that depth. Ground vehicles would have range and endurance characteristics to support operations over something on the order of two hundred kilometers or more. What we have seen over the last few years in Ukraine is consistent with these parameters with UASs of various ranges employed to tactical and operational depths in large numbers. The bulk of the tactical UASs in Ukraine have been small first-person-view (FPV) UASs, which require human interactions and secure communications resistant to jamming. The medium and small UAS concepts described here are more autonomous, are recoverable and reusable, and generally do not require human control.

The concept described would replace current maneuver forces. Additional ground forces would probably still be needed to provide operational depth, including long-range fires and extended-range surveillance. I say probably, because it isn't clear how much of these functions could be off-loaded to space and airborne forces. If they are still necessary, physics would dictate larger platforms and longer ranges for the components of the extended-range suite of capabilities. There are strong efficiency and total-force-resiliency arguments for the inclusion of long-range ground-launched munitions and the supporting C3BM and target acquisition capabilities necessary to support them. The challenges to automating these functions are arguably less than for the maneuver and close-combat functions.

The concept would be incomplete without consideration of sustainment capabilities. What I have described is not a throwaway expendable force. It does have an operational virtue that some systems can be sacrificed to find the enemy, to force them to engage, or as part of a deception. These behaviors should be part of the automated and semi-automated suites of tactical behaviors, but the air and ground vehicles I have described are going to be sophisticated enough and costly enough that they should be reused over multiple missions or operations and therefore must be supported. The medium sized UASs and the ground vehicles would be tactically self-deployable, together and separately, but they would need transporters to bring them into an operational area.

If we apply the projectile, launcher, and transporter concept described above, it looks something like this: The line-of-sight projectiles are fairly conventional—small missiles, grenades, or bullets. It's necessary to have the engagement capability for a variety of targets, so suitable mixes of munitions are needed. To the extent that flexibility to carry different projectile types can be designed in, it should be incorporated into the UAS launchers/transporters. The launchers are the UASs and unmanned ground systems (UGSs) in the concept. UGSs also serve as tactical transporters for the UASs. Each UAS should be able to carry one to ten stowed munitions depending on the UAS and the target type. The air vehicles will be transported on the ground vehicles. Depending on their form of propulsion (battery, liquid fuel, or fuel cell) and weapons load, they will need to have their energy source replenished and potentially weapons replenished from the ground vehicles. The ground vehicles will deliver themselves forward after transport by a larger class of system in some non-tactical mode—air, rail, road, sea, or a combination thereof. Ground vehicles could be resupplied forward autonomously or self-recover to a more centralized refueling and rearming point. There would be no organic maintenance capability. Mission-incapable ground vehicles would self-evacuate, if able, or be towed from the operational area by other ground vehicles if salvageable. In some cases, air vehicles could also self-evacuate. As in other areas, there are trade studies that would have to be conducted to optimize the level of investment in repair capacity and where or if it should occur, but repairing forward is not intended.

Human command and control (C2) would begin at a level roughly equivalent to battalion or possibly brigade. Lower-echelon engagements would be monitored at this level but not tightly controlled unless by exception. The organizational concept is intended to be largely self-sufficient. The tactical echelons described can conduct ISR locally, plan maneuver and engagements, and manage operations as they unfold. Provision will be made to provide offboard support, including intelligence, targeting, and supporting fires from higher echelons and other domains, but outside support is not a critical operational dependency. C2 would be distributed, with each echelon's UGS capable of managing the units or platforms currently assigned, and a small group of UASs acting as a team would self-manage as a team. Dedicated manned C2 vehicles would be introduced at something comparable to battalion or brigade level. These vehicles would be designed for improved survivability based on deception and concealment primarily, and they would be able to operate several kilometers away from the balance of the automated force. Organic airborne communication relay systems using dedicated UASs are likely to be required as a result.

BUILDING BLOCKS

Small and medium UASs will take off and land vertically from UGVs and have operational ranges and endurances consistent with missions of one to four hours and payloads of ten to one hundred pounds for munitions (one of many trade studies needed to optimize the concept) in addition to organic sensing, navigation, computing, and communication capacity. An alternative worth analyzing is to distinguish ISR UASs from weapon carriers. At a minimum, the idea of using some common UASs with no weapon loadout as dedicated ISR platforms for a given mission should be considered. Projectiles carried by and launched from UASs would be relatively short range to reduce weight and cost. The UASs would be hardened against electromagnetic warfare and cyberthreats and be able to operate in a reasonable range of weather conditions. They would be compatible with ground vehicle transport and servicing for energy replenishment and expendable resupply. Growth capacity for survivability features would be desirable. They will be able to engage crewed helicopters and some UASs in a sacrificial mode of attack at least. UASs would be mass-produced using commercially based components and processes to the extent possible with nominal unit cost goals on the order of $10,000 or less for medium UASs and $1,000 for small UASs.

Autonomous ground vehicles will function as UAS transporters, support vehicles, and indirect-fire launchers. They would provide cost-effective operational movement on the battlefield. They would generally be wheeled with roughly the mass and form factor of a medium-sized pickup truck. They would be mass-produced with modular mission packages, especially UAS transporters and indirect-fire launchers. Tactical indirect-fire variants will carry weapons with medium range (out to, say, ten or twenty kilometers) engagements. All variants may carry a limited self-defense direct-fire system, with small-UAS engagement capability and line-of-sight sensors, generally passive sensors. Although the UASs are the preferred direct-fire engagement systems of the concept, the UGVs may have a tactical capability beyond self-defense. They can be networked locally and programmed for collective tactical action in small numbers. Ideally, the UGSs and UASs should be able to operate together tactically in a combined-arms direct-fire assault or final protection mode, but this would probably not be the norm. The trade space for these platforms balances cost, range, protection levels, weapons capacity, and sensor and communication payloads with a strong emphasis on cost. As the weapons and fuel for the UASs are exhausted, UGS vehicles can autonomously redeploy to rearm and refuel or recharge.

Local information systems or architectures, including communications, sensing, processing, and data storage, are the heart of the concept. Together, they must constitute a balanced design that is resilient and hardened against attack. Processing and data storage should be pushed to the edge of the architecture to minimize the burden on the communication links and support as much platform and lower-echelon-level autonomy as possible. Most architectures designed in DOD today are too centralized and intended to support a heavily human-supervised or human decision-making function. Commercial technologies will be important to this architecture, but military requirements will exceed and be very distinct from commercial capabilities. Collaborative sensing and tactical decision-making based on AI will occur and be automated, but primarily at the tactical edge, empowering individual platforms and small collections of platforms organized virtually to collaborate. Above this level, the focus will be on higher-echelon planning and force management based on reasonable aggregated and processed data provided by lower echelons and external sources.

COMPLEXITIES

Most discussions of future military applications of technology focus on the fighting platforms and weapons, but those are the tip of the iceberg and may not be the most challenging aspects of a new force design. In this section, we focus on a long list of complexities that will impact force design. We start with some fundamental operational needs not fully discussed above, and then move to a range of design requirements that have to be considered, and finally wrap up with other, often noncombat, military functions that a complete force will have to meet. Note that, in general, modern mechanized armies are composed of about 10 percent or 20 percent fighting vehicles, and the balance is trucks and the things they carry. There are good reasons why that is the case. A ground force has to perform many functions and meet many needs. The list that follows is certain to be missing some of those items, but I believe it is a good start.

Fundamental Operational Needs

Command, Control, and Communications: Any unmanned-system-intensive concept is highly dependent on resilient and effective C3. The C3 architecture is the great enabler for all the concepts, but it's especially important for the ground domain due to terrain effects on communication links and the need to form ad hoc operational networks, as well as the sheer numbers of participants, the need to deconflict them, the proximity of jamming threats, and the amount of information that has to move on the network at scale. In the commercial world, 5G is the state of the art and

6G is in early stages of development, and there may be direct application of these technologies to ground operations. Military C3 networks are particularly challenging for several reasons; I don't believe there is a network today that can support the concept I've described in a contested environment. I do believe, however, that such a network is obtainable. Achieving that will require careful design of the information flow the network has to support and of features unique to military systems. There will also have to be well-developed options for reduced performance under stress down to some operationally viable minimum product level. As a design principle, I believe as much processing as possible should be conducted at the tactical edge, generally on platforms that carry surveillance sensors as payloads. The target identification function described above should happen at the sensing platform level as much as possible so that only target classification, confidence level, and tracking information is passed to more centralized nodes for fusion. The DOD has tried in the past to create architectures in which much more information is shared among platforms and passed to higher echelons, often with the goal of real-time human interaction with fused imagery. If we start with that goal, this problem may be intractable. If we start with something much less ambitious, but still operationally transformative, I think we can get there.

Extended-Range Precision Fires, Including Hypersonic Weapons: The concept focuses on relatively short-range engagements and continuous control of the tactical battlespace. It provides for automated longer-range systems, but the cost per kill of those systems is expected to be higher than the UAS-carried munitions, and they don't have the same enduring and robust control over terrain. They do have the ability to mass fires, but the UASs have that capability also, although less responsively. A trade-off to balance these capabilities would be needed. Traditionally, artillery has been the most cost-effective weapon in ground combat because of its flexibility, accuracy, and low latency and cost, but the advent of the inexpensive UASs being used in Ukraine and those described here creates a new balance that should be analyzed. I have seen some reports that in Ukraine, UASs have replaced artillery as the most lethal weapons on the battlefield. Precision munitions (missiles or shells) are relatively expensive but significantly enhance the effectiveness of artillery. The value of ground-launched hypersonic weapons, at least against tactical targets, isn't as clear to me. They have faster time on target but are expensive and may not be able to engage mobile ground targets effectively. The concept does not depend on them, but their cost-effectiveness, at least against high-value targets, should be investigated.

Logistical Support—Expendables and Consumables, Infrastructure, Supply, Repair: Some of this was covered above in the description of the concept. The logistics concept for the whole force should be designed, at least to first order, as a next step in fleshing out the concept I've described. An unsupportable force isn't a force. As indicated earlier, most of the Army, by a wide margin, is trucks, not combat vehicles. There are tens of thousands of trucks in the Army—over one hundred thousand High Mobility Multipurpose Wheeled Vehicles (Humvees) alone and only a few thousand combat vehicles. This concept could fundamentally change that ratio, but there have to be provisions to flow fuel and ammo and the items necessary to support the humans in the concept to where they are needed, both by ground and by air. This could be accomplished largely with automated unmanned systems (possibly commercial platforms), with special automated handling equipment compatible with the manned and unmanned receiving platforms. The dramatic reduction in numbers of humans in the concept should relieve many current logistics requirements, but many functions, including medical, graves registration, personnel replacement processing, detainee and refugee processing, and so on, still have to be considered and provided for in the force, even if for lower numbers.

The logistics support system itself must be resilient enough to function under attack. The Ukraine conflict has already demonstrated how drones can threaten historically relatively secure rear areas. The concept I described above is intended to include automation for as much of the logistics function forward as possible. For example, there would be no organic maintenance personnel; there would be reliance on self- or sister vehicle platform evacuation, and there would be automated rearming and refueling. Some operational attrition of the force due to logistics shortfalls may be acceptable, but it shouldn't be widespread. Logistics cannot be the concept's Achilles' heel. Modular designs would make automated intermediate-level repair in the form of module exchange more tractable. In this and any future concept, AI-based support functions, including proactive prescriptive maintenance/evacuation and inventory management, should be part of the design trade space. As in the kill-chain approach to engagement analysis, the whole logistics chain must be analyzed and optimized as part of the concept.

Multi-Domain Considerations: The ground domain concept described above is built around an operational force that is designed to take and hold ground. It includes UASs designed to support that mission. I didn't let the fact they are aircraft constrain the concept; their role is primarily to find and attack ground targets in support of ground operations. Forces from other domains can also attack ground targets. For the future ground concept, and all others, relevant information to support operations will come from other domains, in this case, primarily air and space systems. Fires can also be provided from other domains, especially air. In addition, support functions, such as some transportation, communications, EW, cyber support, and positioning, navigation, and timing (PNT), will come from other domains. Finally, cross-domain support can surveil, target, and engage threats of high priority to success in the ground domain. Those enemy targets include enemy command and control, surveillance systems in any domain, and especially space- and air-based systems that provide the enemy with long-range precision targeting and fire capacity. The design challenge is to optimize this cross-domain collection of capabilities and to simultaneously make it resilient enough in the face of enemy attack to be successful.

Tactical Mission Variations (Scouting, Hasty and Deliberate Offense or Defense, and Movement to Contact, for Example): These missions would still exist, and the concept has to accommodate them. There is nothing prohibitive about applying the concept to address these and other missions through several essentially templated and automatable tactics, techniques, and procedures (TTPs) associated with each mission. The availability of infinitely tailorable tactical organizations, without regard for unit cohesion or training experience, provides enormous flexibility. This is one place where human decision-making would be useful, at least initially, in defining missions for tailored units of unmanned systems. What should be automated are the decisions about tactical behaviors at the small unit level and individual platform level within an assigned mission. As the concept matures, more sophisticated and higher-level automated tactical and operational decisions should be possible, but that isn't necessary at the outset. Even at early stages, AI can support human decision-making about courses of action.

Target Identification/Collateral Damage Avoidance: This is the core enabling technology for lethal autonomy. It has to be at least as good, and probably demonstrably better, than human decision-making in the same situation. This is a measurable activity that can be tested in simulation and in the field. It is also a capability that can be traded operationally in real time by adjusting decision algorithms to improve performance or react to changed conditions (the appearance of a large number of civilian refugees in an area, for example, or conversely, the indications of a major enemy advance through an area). Human monitoring and override authority would be part of the concept. In less intense environments with more time available for decision-making, human control of individual engagements could be exercised.

Design Requirements and Considerations

Civilian Engagement and Interaction: It will happen, and the autonomous behaviors, especially by the ground vehicles in the concept, will have to accommodate the range of interactions. The range includes friendly, neutral, and adversarial, as well as both individuals and groups. When the force is operating among potentially hostile civilians, provisions may have to be designed in to affect a range of responses, and there would be a way to communicate with civilians to reduce risk and manage any given situation. Moving—or not—through hostile protesters comes to mind. What to do about children throwing grenades comes to mind. This is one place where human remote involvement or control might be necessary. The default in some cases would be to avoid injury to humans completely, and in others to sacrifice the vehicle or give a lethal response. What it may come down to is that the force defined in this ground-domain concept and designed to defeat a conventional armed ground force at scale isn't suited (at least not without human augmentation) to the human-intensive environment of a counterinsurgency or counterterrorism mission. That said, the elements of this concept could still be very useful in such an environment.

Confidence in Automated Behaviors: It can never be perfect, but several constraints can be in place to prevent mistakes. These would include temporal and spatial boundaries on movement and/or fires, the requirement to seek human permission if certain thresholds are reached (such as the numbers of engagements, numbers of targets identified, or the identification of possible cyber or EW actions). To improve performance and confidence over time, testing and machine learning opportunities should be maximized over the life of the systems. Updates to automated decision tools should also be almost continuously issued at scale.

Countermeasures for Self-Protection: Countermeasures designed to defeat incoming attacks will have to earn their way onto the platforms based on their cost-effectiveness. Current examples include active self-defense systems on armored vehicles.

Deception: The US has always underemphasized deception and the use of decoys or deceptive countermeasures.[3] Concealing the manned C2 vehicles should be very high priority. Part of this design challenge could be making them indistinguishable from lower-priority military vehicles or even from commercial vehicles. The UGVs are relatively high-value targets in this concept; they carry multiple "archers" that carry multiple

---

3 One DARPA reviewer suggested making deception a warfare domain in itself. I wouldn't go this far, but it does need to be elevated in importance, in all domains.

"arrows." Decoy UGSs could be very cheap and cost-effective. They even open the possibility of anti-simulation, where some decoys can be less decoy-like and give the enhanced appearance of being the real targets—creating a false confidence that the real targets have been identified and engaged. The UASs would rely on low-level flight and concealment as well as favorable cost-exchange ratios with some threats. Decoys could be used to exhaust defensive munitions as a precursor tactic. The potential for operational-level deception is also high with this concept and should be included during course-of-action planning and analysis.

Default Behaviors and Loss of Contact by Unit or Echelon: One operational contingency that would be of concern to the operational community is the implications of lack of contact or control over a platform or collection of platforms, which, in this concept, would include UGVs and UASs and ad hoc organizations composed of them. There would have to be programmed abilities to identify when this had happened and embedded tailorable rules at each echelon for how to best respond. There is no one optimal solution to this question; the "right" answer could range from "continue the mission" to "withdraw all units immediately," depending on the tactical situation. Setting these rules in advance is another human executive role that would likely be needed in the initial fielding of the concept. Because this can happen to any unmanned platform or group of platforms, it gets a little complicated, but it isn't unmanageable, with different options for each possibility of concern or class of possibilities.

Energy: Current ground forces live on hydrocarbon-based fuel consumed by internal combustion engines. This includes fuel for vehicles and for generators for all supporting units. The potential exists for alternative forms of fuel for some applications, but I don't see any way to eliminate the need to refuel operational vehicles, both the UGVs and the UAVs as well as all the supporting vehicles. Even if, for example, fuel cells or all-electric vehicles become cost-effective and operationally viable, they will still need to be refueled periodically (in some form). The frequency of refueling may be much less. Current armored vehicles must be refueled almost daily. A combination of much lighter-weight platforms and more efficient technology should substantially reduce the operational energy burden, but the force will have to include the infrastructure to make some form of refueling possible. As noted elsewhere, that infrastructure, including the fuel distribution and transfer systems, will be highly automated.

Humans in the Concept: The location, role, and support of humans is discussed under other areas and in the concept description. The bottom line is we must have some humans in the concept. Even though the numbers are reduced significantly, we still have to make provisions to support those humans, which will include C2 operators, special operators, and some logisticians. The survivability of these humans would be a high priority. It would be achieved primarily by concealment, deception, and keeping them away from engagements. Both the Ukraine and Nagorno-Karabakh conflicts have already demonstrated the need for extensive personnel protection measures. Another notable feature of the concept is that the military branch system (armor, infantry, artillery and so on) may disappear or be significantly changed. The branch system is based on the need for humans with highly specialized knowledge and training about some fraction of the force and certain equipment and units. In this concept, the focus for humans is much more on integrated operations and the ability to supervise those operations and interact with AI systems and automation—a more generalized and technical but highly specialized skill set. In addition to designing the concept to support and sustain the humans in the system, C3BM systems (including AI-based tools) will have to be designed to facilitate human interaction and support human decision-making where it is needed or required.

Infantry, or the Lack Thereof: The concept notably doesn't have infantry units or mechanized infantry units as its centerpiece. That's because it is designed to fight a conventional force at scale and doesn't need them for that purpose. That opposing force would include mounted and/or dismounted infantry, but the concept is for the UASs, supported by the UGVs and long-range precision fires, to make massing and movement of enemy vehicles and dismounted infantry untenable. Without vehicles, human infantry is slow, limited in carrying capacity, and needs extensive sustainment support. The concept assumes enemy infantry could be essentially fixed in place by the UASs and defeated over time, even if stationary concealment were temporarily successful. The infantry threat would be placed under continuous surveillance and covered by fires. It would be destroyed if it moved or exposed itself in an attempt to conduct engagements.[4] As a result, neither infantry nor a robot-like unmanned platform with mobility characteristics similar to humans is required in the concept for the driving force-on-force mission. For situations where human contact with civilians is inherent in the mission—peacekeeping, counterterrorism, and some counterinsurgency missions—armed humans will be part of the equation, but even in these environments, it will be preferable to put unmanned systems rather than humans in harm's way whenever possible.

Interoperability: For the US in particular, it is a strategic requirement to be able to work with allies and coalition members. In the past, this has been given some attention in peacetime, but generally underemphasized, resulting in jury-rigged work-around approaches. The envisioned ground forces are likely to deploy into a situation where US forces must operate in proximity to and in cooperation with allied units. Interactions at all levels, individual systems, small units, and operational levels are going to happen. In addition to combined operational planning and force management, provisions to ensure the safety of allied forces and to prevent engagement of US forces by friendly allies will be required. Given the pace at which the concept I've described would conduct operations, and the supervisory role of humans, a lot could go wrong in a hurry. Cooperative designs and shared identification features that enable allied unit integration and build confidence in unmanned systems and collective behaviors are a necessity.

Legal Design Constraints: The lawyers are always with us. All our autonomous systems and the units mostly composed of them must fol-

---

4   Vikram Mittal, "Ukraine's Battlefield Defense Is Forcing Russia to Adapt Its Tactics," *Forbes*, January 16, 2026, https://www.forbes.com/sites/vikrammittal/2026/01/16/ ukraines-battlefield-defense-is-forcing-russia-to-adapt-its-tactics/.

low the law, including the law of armed conflict.[5] There are also issues of liability for accidental damage to property and person, but these should be manageable under civil law approaches. Programmed behaviors and decision processes must take these constraints into account, but the operational concept won't work if legal constraints and decision-making have to involve humans in every situation. I've had the chance to observe, with admiration, how carefully legal concerns are weighed and evaluated in counterinsurgency and counterterrorism operations. We can tailor how this is done to the tactical situation, but in high-intensity force-on-force ground operations, there is no time for legal review of all tactical decisions. As a result, the automated rules of engagement must incorporate legal constraints effectively. For individual engagement decisions and small-unit behaviors, they must be part of the automated functionality. For mission-tasking decisions, they must also be part of the automated process, but where appropriate, provisions should be included for human judgment and intervention. A good example is an assault or coordinated strike on a village where an enemy force is present but might be using human shields or where humans might simply be present. In another situation, a similar strike or attack against an enemy unit in an assembly area, or moving to contact in low-civilian-density terrain, would accommodate a much more automated approach. In even these situations, the combination of AI and human judgment may give a superior result. Consider all the devastating mistakes made in Afghanistan and Iraq with mistaken strikes on weddings and hospitals, even with the best of intentions. AI-supported decisions should improve on this performance. The Geneva Conventions, including the legal obligations regarding prisoners and wounded prisoners (see below), must also be considered.

---

5   The second Trump Administration seems to have deemphasized compliance
     with the law. Secretary Hegseth has spoken about "maximum lethality, not
     tepid legality." My assumption is that future administrations will restore
     compliance with the law as an important professional norm of our military.
     It is in our interest to do so if we wish to maintain the professionalism of our
     military and to hold others accountable for violations, as we have in the past.

Physical Security: For UGV systems, the designs must include some provision for vehicle and unit security from local threats. Some UASs, small and medium, can be employed for this purpose. Modest hardening can provide a level of protection, but self-defense capability is also needed. A nonlethal self-defense mechanism should be considered as a desirable part of the ground vehicles' suite of equipment, but one needs to be careful throughout the design about requirements creep. The UASs could include anti-tamper and self-destruct features as desired.

Position, Navigation, and Timing: This service is essential to any domain, including land. GPS, as it currently exists, is not secure enough to provide the required level of resilience. There are a few technologies and combinations of technologies in various stages of fielding and development, but adequately resilient PNT must be part of the concept. Distributed autonomy, data fusion, and sound decision-making of all types are totally dependent on PNT availability. This problem has to be solved for any future concept, highly autonomous or otherwise.

Prisoners of War: People have tried to surrender to drones in the past. In the concept I've described, that could well happen. Both UAVs and UGVs would have to be programmed to recognize an attempted surrender and respond correctly when that occurs. For UGVs at least, there would need to be an automated or semi-automated mechanism to accept a surrender and escort prisoners to a collection point, if circumstances permitted. Alternatively, prisoners could be kept in place under guard until processed. Wounded prisoners couldn't be treated but they could be allowed to assist each other. Rules of engagement would also have to anticipate situations where prisoners try to escape or renege on their surrender. This is a complex issue, especially for dealing with an enemy who tries to surrender while under fire. We deal with it already, however. Attack helicopter pilots, for example, can't accept a surrender. They must decide whether to stop the engagement or continue. The unmanned machines in the concept will have to be programmed to do the same.[6]

---

6   In the classic *Men Against Fire* (William Morrow & Company, 1950), S. L. A. Marshall noted that experienced combat infantrymen in World War II knew that if they

Reliability: This was addressed earlier in the discussion of models for sustainment. The platforms in the concept and their major subsystems must be designed with adequate reliability to function operationally long enough and well enough to be cost-effective. High reliability, such as in long-endurance spacecraft, can be very expensive, but that isn't required here. The absence of humans to provide organic unit-level maintenance of the systems forces a design for reliability trade-off that ensures adequate confidence the concept will perform as needed at the aggregate operational level. It would be a mistake, as it has been for many military systems, to demand more reliability than is necessary or cost-effective. It would also be a mistake to under-specify reliability parameters. Anticipatory failure and fault monitoring to permit unmanned system self-evacuation prior to failure is also likely to be cost-effective for some subsystems and components.

Requirements Creep: Conscious attention should be placed on avoiding cost imposing marginal or low return-on-investment requirements. The natural tendency in all domains is to add more requirements to the design until the concept crashes from its own weight. A lot of the items on this list are good examples of potential requirements creep. The whole point of the concept is improved cost-exchange ratios over current systems.[7]

Responsive Threats: The first line of defense against the ground domain concept would be quick-reaction counter-UAS systems and concealment, both of which we've been seeing extensively in Ukraine for some time. In Ukraine both sides move quickly to respond to developments by the other, especially the proliferation of battlefield frontline ISR and lethal

---

tried to surrender during a firefight, they would be shot. That environment doesn't permit split-second decisions to stop shooting because someone puts their hands up and exposes themselves. The way to surrender with a chance of survival was to hide until the shooting stopped and then indicate your presence and intention.

7   This item will occur in every domain. My apologies to the reader for the redundancy, but I've seen too many systems fail because of requirements overload. The Navy's recent frigate-class ship cancellation is a current example.

drones.[8] Based on DARPA experimentation to date as well as the Ukraine experience, we can expect some novel attempts to defeat AI-based detection and identification. The next level of response would be attempts to design dedicated countermeasures exploiting weaknesses in the new threat. Commercially derived drones have been especially vulnerable to countermeasures that they were not designed for. For the UASs and UGSs in this concept, we can expect extensive EW and cyber responses and dedicated counter-UAS weapons. Finally, we would see attempts to design and field forces like those described here, but superior in performance (longer range, more hardening, greater standoff, more expendable, and so on) in areas deemed by the adversary to be operationally important. Since I wrote the original version of this book, we have seen this exact scenario play out in Ukraine.[9]

Safety: This is always a design consideration. Some aspects of safety are covered in other sections, but unmanned autonomous systems will pose some unique problems, and the very limited presence of humans in the concept may add unique safety concerns. Without human operators, any number of risks that humans could have mitigated will have to be designed out or addressed through automation. The list of safety concerns includes electromagnetic radiation, chemical, electrical, mechanical, and explosive risks, among others. It also includes all possible storage, transportation, or operating conditions the autonomous vehicles will ever encounter, in war and peace.

Single Points of Failure: A detailed design would have to carefully consider critical communication links and nodes and provide some com-

---

8   For a fascinating comparison of Ukraine's and Russia's approach to this problem, Mark Krutov, Sergei Dobrynin, and Yauhen Lehalau, "Inside Rubicon, the Elite Russian Drone Unit Wreaking Havoc on Ukraine's Troops," *Radio Free Europe/Radio Liberty*, September 17, 2025,https://www.rferl.org/a/russia-drone-rubicon-secret-ukraine-war/33532804.html.

9   To the extent that the US is trying to acquire inexpensive counter UAS systems developed by Ukraine to use in the war against Iran. See: *These are Ukraine's $1,000 interceptor drones the Pentagon wants to buy*, *Military Times*, March 11, 2026, at https://www.militarytimes.com/news/pentagon-congress/2026/03/11/these-are-ukraines-1000-interceptor-drones-the-pentagon-wants-to-buy/.

bination of resiliency and redundancy where necessary. I don't see any obvious major problems in the concept from this perspective, but it has to be addressed. A thinking and capable opponent will certainly look for weaknesses in the concept to exploit. PNT, mentioned above, is sure to be one area of interest. Others might include the ability to overwhelm the concept's tactical information-processing capacity with an operational "denial of service" approach where the numbers of potential objects to be processed is overwhelming. Embedded cyber vulnerabilities are certainly on this list also.

Stealth: Some degree of stealth is important for the ground vehicles in the concept. While I assumed a conventional human-intensive mechanized and/or infantry threat as the initial opponent, since I wrote the original report five years ago it has become very clear that one has to assume that threat will include large numbers of UASs that can attack ground vehicles (Ukraine, Nagorno-Karabakh, and ISIS all have demonstrated this). It isn't practical to provide top and all-around hardening of ground combat vehicles against those threats, today or in the foreseeable future. The first line of defense, because of its cost-effectiveness, should be concealment through practical levels of stealth or camouflage. Signature management might be a better description. Part of the suite of passive defense measures could be movement patterns, use of terrain for concealment, and signature management, in some cost-effective combination. UGS movement would be episodic and could be structured to be hard to correlate to traditional military unit movement. For UASs, signature management to some level should be explored as a design trade. Countering enemy surveillance systems to prevent targeting of all these vehicles, and especially the manned command and control vehicles, would be a high priority.

Terrain Variations (Forest, Urban, Built-Up, Desert, Mountain, Swamp): The UGV envisioned should be able to operate in all terrains, with some limitations in each. Wheeled or even tracked vehicles with the payload capacity described in the concept have some fundamental design limitations. Because the UASs are the primary targeting and engagement asset and are not generally constrained by terrain, the biggest job of the UGV is to get to where the UASs can be effective. This relaxes the requirements for those vehicles considerably. The UASs could cover terrain inaccessible to the ground vehicles by observation and fires. To penetrate and secure structures, the concept would have to include smaller unmanned systems that could operate in buildings. These could be transported by the UGV and deployed directly from the UGV or by the UAS. If the terrain cannot accommodate the UGV, it cannot accommodate a mounted enemy force either. Dismounted enemy infantry threats will be discussed below.

Training, Experimentation, and Testing: The concept opens up some very cost-effective opportunities to design for training the AI included in the concept, the humans that are part of the force, and the two together. The efficiencies come about because the human role is largely remote from the physical world in which engagements and tactical decisions are being made. This permits very realistic simulation and exercise construction in a virtual environment identical to the one that would be seen operationally. The human role is supervisory and much less physical than in current forces, largely eliminating the need for traditional field training. Training, experimentation, testing, and upgrades merge in this concept and would be all but continuous. Any operational experience should immediately inform both humans and the AI embedded throughout the concept.

Transportation: This was mentioned briefly in discussing the concept. There have to be assets that can get the combat and supporting systems and units to the battlefield. As today, everything in the land domain concept has to be designed to permit movement by a range of commercial and military means. In the future, many of those transportation platforms will also be unmanned. As a result, the vehicles in the concept will probably have to be transportable in a combat-ready configuration (fuel and ammo on board), or close to it. Operational movement of the force could normally be organic, but some provision for at least intermediate operational airborne movement, by rotary wing or tilt-rotor aircraft, would be needed to provide operational flexibility and agility, a form of vertical envelopment. These transportation systems would be unmanned and could serve several purposes: evacuation, resupply, and operational movement.

Unattended Ground Sensors: There would be a potential role for these, and at times, the UASs could land and take on this role on a temporary basis. They or the ground vehicles could also deliver sensors and emplace them. In some cases, unattended smart munitions could be included, with options for human control depending on the situation and in accordance with US policy (see the mine warfare discussion below). These systems could be recoverable or expendable.

Vehicle Recovery: UGVs and UGSs will self-recover as much as possible. If cost-effective, at least some UGVs should have the capacity to autonomously connect with and tow a similar-sized UGV to a recovery area. Highly damaged UGVs or those in complex recovery positions could be abandoned or recovered by special purpose vehicles. UASs may be recoverable by either UGSs or special purpose vehicles, but it isn't obvious what the cost-effective requirement would be. All vehicles should have some ability to auto-destruct or otherwise neutralize sensitive data and equipment.

Weather Implications: The concept must be designed to operate in a range of weather conditions, but not the most severe transient extremes. High winds and some visibility conditions could limit the UASs in the concept. The intent would be to adopt a more defensive posture during these periods. The insensitivity of unmanned systems to cold, heat, and humidity relative to humans would be an advantage.

Other Needed Military Functions

Air Defense: The biggest air threat to this force is small UASs operated by the enemy, as we are seeing in Ukraine. These systems are already being fielded and used at scale, so this isn't a future hypothetical threat. Each ground platform would have some organic self-defense capability. This is likely to be a gun system that would also provide self-defense against ground targets. A defensive suite may also include soft-kill capacity (jamming). Specialized unmanned ground platforms with directed energy systems are possible and would have to be distributed within the force.

These platforms would be unmanned and would include soft-kill (EW and cyber) capacity as well as kinetic kill. The UASs in the concept could have autonomous counter-air capability as well, if necessary, with the possibility of dedicated counter-air munitions. Enemy launchers of UAS threats would be high-priority ground targets. For both air and ground platforms, cooperative and noncooperative Identification Friend or Foe (IFF) provisions might have to be included to preclude fratricide.[10]

Civil Affairs and Psychological Warfare: These human-intensive or human-centric functions are not the focus of this book, but they would be considerations in any large-scale land operation, both those involving conventional forces and those associated with irregular warfare. The enemy's will to resist and support for, or at least acquiescence to, an occupying force are central to the overall success of a campaign. The introduction of forces structured like the ones I describe, with high reliance on autonomous weapons, opens some interesting psychological warfare opportunities. Any time human-occupied areas are seized from an enemy, there will be a responsibility to care for the population of those areas and restore civil government functions. It's beyond the scope of this book, but collecting, controlling, and managing information and the flow of information to the enemy military and civilian population should be a key element of any operational campaign. Much of the work on commercial artificial intelligence today addresses the problem of persuading civilians to take certain actions—generally buying something, but increasingly to influence political opinions. That technology should be directly applicable to these functions. In addition, the technologies associated with tracking the activities of entire populations through their interaction with the digital universe have significant utility here. It's a brave new world for civil affairs and psychological operations.

Combat Engineer Support: Some specialized units/platforms/mission modules for obstacle removal or creation, bridging, and demoli-

---

10   A lot of effort has been put into countering small UAS systems ever since
     ISIS started using them at scale over ten years ago. This effort continues
     without any cost-effective silver bullet having been found so far.

tion at least would be needed, as would some construction capacity, but the intent would be to minimize the need for highly specialized vehicles and units. Specialized platforms like bulldozers, breaching systems, and bridging systems that might be used under fire would be unmanned and either autonomous or operated remotely. I don't see a high value or likelihood in the future of US forces needing or encountering highly fortified positions that would have to be assaulted. If they were encountered, the forces in them could, in most if not all cases, be fixed in place and isolated until they surrendered and/or were neutralized by fires.

Cyber and EW: Both are enduring threats and opportunities that require continuous attention in any concept and in any domain. By putting as much decision-making capability into each autonomous platform as practical, the concept reduces the traffic required to the minimum. Cyber hardening has to be part of every design and continuously updated. EW hardening is necessary as well and must be continuously updated. We need new paradigms and state-of-the-art capabilities in these areas no matter what. Offensive cyber and EW are highly specialized and should be designed into the concept, probably in a mix of dedicated specialized platforms (manned and unmanned) and in some distributed capability, especially for EW, as a secondary mission for all platforms. Software-defined radio frequency (RF) systems open areas for integration of communications, sensing, electronic warfare, and cyber to explore and incorporate in all future systems.

Disaster Relief/Humanitarian Assistance: I don't see this as a core mission for this concept. Some assets, such as transportation, could be useful and should be applied. Most current ground force humanitarian assistance is based on the large-scale human support capacity in the military. This concept is not human-intensive, so capacity to provide medical, shelter, power generation, and subsistence support wouldn't be available in this force with the same density or volume as current forces. The ground force should certainly provide whatever it could, but that would be limited by the nature of the new concept and the much-reduced combat service and support structure it entails.

Intelligence Integration: The concept must include automated tactical intelligence integration to support operational and tactical-level planning, including course-of-action analysis and the generation or update of the equivalent of an operations plan. Intelligence and operations are fully integrated in this concept, and there is no significant lag between receipt of new intelligence data, operational planning updates, and mission execution. This elevates the pattern recognition problem to a higher level, where the decisions being automated or supported are about unit planning at both the tactical small-unit level and higher, as opposed to about selecting targets for engagement. This is where successful integration of data from multiple sources, including some in other domains (notably space and air in this case), becomes important and where humans are still likely to retain a role in command decisions.

Mine Warfare: Traditional mines are not very effective weapons; while individually cheap, they have poor aggregate exchange and cost-per-kill ratios. They also are expended in use and leave behind devastating long-term hazards to civilians. Their biggest role is slowing or channeling an enemy advance to increase enemy exposure to fires. Smart land mines can be more effective, but at a higher cost and, under current policy, with a heavy burden in human control. The US has appropriately all but eliminated traditional emplaced or scatterable mines because of the residual danger to noncombatants.[11] The emotional effect mines have on humans advancing is, of course, lost when the advancing force is automated and unmanned. For the future force I have described, advancing against minefields can be accomplished in part through avoidance, particularly where the armed UASs are carrying the bulk of the fight, and if unavoidable, through the sacrifice of some UGVs (purposely configured to be more expendable perhaps). Specialized mine-breaching equipment is possible

---

11   On December 19, 2025, it was reported that Secretary Hegseth was reversing long-standing US policy on the use of land mines. Joyce Lee, "Hegseth Reverses Land Mine Policy to Allow Use of Controversial Weapon," *Washington Post*, December 19, 2025, https://www.msn.com/en-us/news/world/hegseth-reverses-land-mine-policy-to-allow-use-of-controversial-weapon/ar-AA1SHlli. I don't expect this change to be enduring.

but may not be needed in the concept. I do not envision US forces needing to employ mines in the future concept.

Operational Planning and Rehearsal: At the platform and small-unit level, planning would be automated and there would be simulated rehearsals, both with human monitoring and supervision, when practical. At higher levels, planning would still be highly automated but with more human interaction and feedback. The cycle of immediate, mid-term, and longer-term planning—anchored by simulation—would probably still occur but be much more compressed, automated, and integrated.

Rear Area Security: A fluid battlespace is envisioned for the concept, but there would still be areas of control for each belligerent, if not a well-defined front line. If there was concern about leave-behind enemy forces or partisans or a rear area raid of some kind, then some tactical forces could be allocated to rear area security. Some logistics systems and support areas would need self-defense capability, which could largely be provided by organic unmanned systems. I would envision tighter controls on any automated engagements in the rear area, up to and including human supervision of each automated engagement.

Special Operations: Ground special operations are currently particularly human-intensive. Some of these missions, destructive raids, for example, could be carried out by autonomous systems. Others might require humans for the foreseeable future—think of the effort to capture Osama bin Laden, for example, or security assistance missions more generally. I don't see the need for some traditional human special operations forces going away anytime soon, but unmanned autonomous systems should certainly be integrated into those operations to improve performance and reduce their risk.[12]

Underground Combat (Tunnels): If this is considered a priority, specialized equipment may be needed. The smaller UGVs and UAVs envisioned for use in buildings should be of utility in tunnels as well. Tunnels are a problem that exists in the context of permanent border situations

---

12   Another more recent example is the seizure of President Maduro and his wife.

or in attacking forces that have had a long time to prepare defenses in situations like urban areas, such as in Gaza. I didn't consider this a major driver for the land domain concept, but it can't be entirely ignored.

CONCLUSION

As in other domains, the concept I have described is transformational. It won't be achieved overnight or in one giant leap. It does assume the obsolescence of the current mechanized force, which will certainly be disputed. Ukraine, however, is providing ample evidence that change in this direction must and will occur. Experimentation, force-on-force virtual and live testing, and more detailed analysis of transformed or partially transformed force designs will be needed to build confidence and mature the needed technologies. That journey has already started.

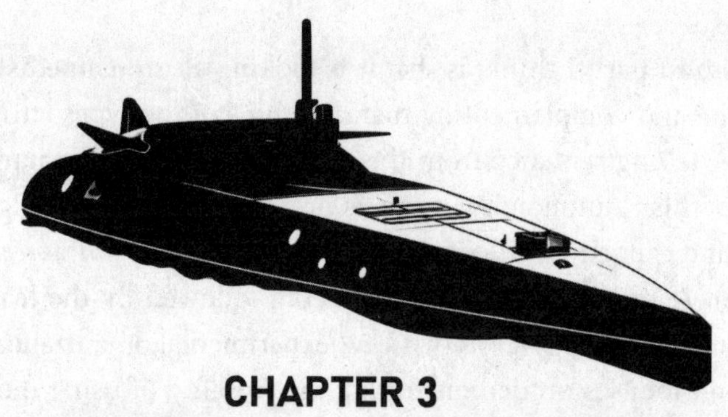

CHAPTER 3

WARFARE DOMAINS– SEA SURFACE

It's been an interesting exercise to watch the Navy struggle with what to do with unmanned surface ships. Assistant Secretary of the Navy Sean Stackley and I were discussing this topic several years ago, and he asked me, I think appropriately, "What mission would they (unmanned ships) perform?" It wasn't a rhetorical question. Neither of us thought at the time that the Navy was anywhere near turning an existing mission over to unmanned vessels (with the exception of mine hunting and clearing), and until someone answered that question, there wasn't a lot of point in buying several unmanned ships, which was being advocated at the time.

The Navy, I'm afraid, still hasn't answered the question, although it is continuing with the acquisition of modest numbers of experimental unmanned surface ships of various sizes. It has also created organizations dedicated to unmanned surface vessel experimentation. In March 2021, the Navy released an overall unmanned "framework" document.[1] The

---

1   My experience with strategy documents of this type coming out of the Pentagon's consensus-oriented bureaucracy is that they are aspirational as opposed to firm, vague in the extreme, and rarely provide meaningful plans or commitments. Department of the Navy, *Unmanned Campaign Framework*, March 16, 2021, https://nps.edu/documents/115559645/121916825/2021+Dist+A+Navy+and+

problem in part, I think, is that it is looking at unmanned ships as augmenting and complementing manned ships of the types currently in the fleet. A telling sentence from the introduction to the unmanned framework is this: "Autonomous systems provide additional warfighting capability and capacity to *augment our traditional combatant force*" (emphasis added). A more recent approach, being followed by the leader of Task Force 66, one of the Navy's two UAS experimental organizations, goes further and focuses on demonstrating the viability of using multiple UASs to duplicate and replace the functions of a destroyer at much lower cost.[2] Most recently, the Chief of Naval Operations has talked about integrating unmanned vessels into the fleet operationally, but still as part of formations dominated by crewed vessels.[3] There's still another way to approach the operational use of unmanned surface vessels (USVs). It begins with freeing oneself from current concepts and paradigms entirely.

OPERATIONAL CONCEPT

As we did for land systems, let's start with a clean sheet of paper and expand our thinking. The reason we have manned surface warships is to provide sea control, protect shipping lanes or at least the ships in them, and project power. This includes the ability to protect international freedom of the seas and commerce, defeat other countries' maritime assets (surface and subsurface), attack enemy land targets in support of air and land campaigns, support our allies, protect the movement of land forces and freight to overseas theaters, and support amphibious operations. Naval vessels also

---

Marine+Corps+Unmanned+Campaign+Framework+16+Mar+2021. pdf/bc9f2eae-f63d-118b-593a-83b4fb2e5211.

2   Lauren C. Williams, "How Many Drones Does It Take to Replace a Destroyer?" *Defense One*, August 27, 2025, https://www.defenseone.com/defense-systems/2025/08/ how-many-drones-does-it-take-replace-destroyer/407743/.

3   "CNO Outlines Vision for Unmanned Systems in Navy's New 'Hedge Strategy'," *Defense Scoop*, January 27, 2026, https://defensescoop. com/2026/01/27/navy-hedge-strategy-adm-caudle-unmanned-systems/.

perform a "show the flag" mission to communicate to others the facts of US presence and power, and to support diplomatic efforts. We should think about three questions, each of which would provoke a strong reaction from any US Navy Surface Warfare Officer.

First, do we need surface ships at all to perform these tasks? With foreseeable systems and technology, it is probably more efficient to perform a large fraction of the sea control mission from places other than the surface of the sea. For surveillance and targeting purposes, airplanes and satellites have much shorter movement times to an area of interest or, in the case of on-station geosynchronous satellites, none at all. Those systems can see much wider areas than surface ships. If sea-based surveillance is considered necessary, a large deployment of unmanned monitoring stations would be trivial in cost compared to manned warships. The seas are already being monitored from a range of satellites and sea buoys as well as from commercial shipping. Some years ago, I had one of my staff do a back-of-the-envelope calculation on how many unmanned wind-propelled and solar-powered floating sensors would be needed to continuously monitor the surface of the Western Pacific. He calculated that the whole job could be done with one thousand small autonomous platforms at $1 million apiece, or for about $1 billion, much less than the cost of a single warship.[4] These assets could be attacked, of course, but that would be a hostile act and the lost assets would be relatively inexpensive to replace. If delivery of weapons is the goal, there may be more efficient platforms than surface warships. Warships are largely launcher, sensor, and projectile magazine transporters. Shore-based bombers and land-based conventional cruise or hypersonic missiles can achieve much faster time on target, unless the warship is already in fairly close proximity to the

---

4   The concept was based on larger-scale versions of products then produced by a nontraditional defense contractor called Saildrone, which received some funding in the early days of the Defense Innovation Unit Experimental in about 2015. The company has significantly expanded its product range since that time. "Saildrone," Saildrone Inc., https://www.saildrone.com/.

targets of interest.[5] While surface ships can have fairly large magazines, air assets can also be reloaded and can complete multiple sorties in a fraction of the time it takes a manned warship to return to port, be rearmed, and move back on station.[6] Spacecraft like the SpaceX Starship could conceivably carry large quantities of munitions anywhere on the planet in time frames measured in minutes. Aircraft carriers and surface ships can stay on station for long periods, but carriers must be replenished with aviation fuel and munitions frequently, and those auxiliary logistics assets are vulnerable to attack. Bottom line, it may be much more efficient to control the surface of the sea from space, the air, and land than from surface ships— if adequate situation awareness can be provided by some other means.

Second, if ships are cost-effective for some tasks, do we need people on those ships to perform the tasks? If one thinks of the ship as a black box that delivers certain capabilities—air defense, surface and subsurface attack, and land attack—there is no obvious reason why those black boxes have to contain people. A surface ship is a combination launcher and transporter with a suite of sensors, self-protection devices, and C3 capabilities. The people are on it to control those functions and take care of the ship. If we can reliably automate or remotely control those functions and the ship doesn't need people to take care of it, there's no compelling reason to have people on board.[7] The US Navy decades ago moved toward substantially reduced manning on newer designed ships. This is enabled by increased automation, system integration, and high reliability. Unmanned autonomous surface ships have been built to various scales and performed successfully on extended voyages.[8] More are under con-

---

5   The laws of physics tend to limit surface combatants to speeds
    on the order of thirty knots. Planing vessels can go much
    faster, but have serious problems in high sea states.

6   Current US Navy ships cannot have their vertical launch cells reloaded
    at sea. Attempts to remedy this shortfall have been unsuccessful.

7   A traditional reason for large crews, coming out of the World
    War II experience, is the need for people for damage control,
    especially firefighting, if a ship is hit by enemy fire.

8   Two unmanned Navy ships transited the Pacific in 2023. Micah Hanks, "US Navy
    Unmanned Surface Vessels Complete First Pacific Journey Amidst Growing Tensions,"

struction. Small solar and wave and/or sail-powered unmanned vessels have also been built and used fairly extensively. Commercial shipping is minimally manned today and experimenting with unmanned concepts.[9]

Finally, there is the serious question of surface ship survivability, independent of whether a ship is manned or not. One of the most significant unknowns in modern warfare is the survivability of warships against state-of-the-art threats, especially anti-ship missiles. Modern warships have fairly capable defenses, but once a capital ship is targeted today, it is likely to be attacked with large raids of sophisticated anti-ship missiles designed to work together to penetrate the ship's defenses and the integrated defenses of any formation of ships. Layered defenses, remote launch, cooperative defense, and a range of soft-kill mechanisms and deception can all be deployed against the attack, but as a former US Army Air Defense Officer and a systems engineer who worked for years on tactical and strategic air defenses, my money is on the attacker. The numbers game favors an attacker; even theoretical layered, statistically independent, high single-shot probability of kill defense systems can be overwhelmed. Only the attacker knows for certain what creative tactics, techniques, technology, and attack strategy they will employ. Probabilities of successful engagement have to compound over many attackers, and defensive missiles can be simply exhausted.[10] A major problem today is that no one knows with any accuracy just how effective shipboard defenses would be against a modern threat. There have been some limited engagements, but there has never been a many-on-many engagement of current state-of-the-art anti-ship missiles in a coordinated attack on a modern warship or formation

---

The Debrief, September 21, 2023, https://thedebrief.org/u-s-navy-unmanned-surface-vessels-complete-first-pacific-journey-amidst-growing-tensions/.

9   The *Yara Birkeland*, an autonomous Norwegian freighter, is currently undergoing experimental use while technical and regulatory issues are being addressed. "Yara Birkeland," Yara International, https://www.yara.com/knowledge-grows/game-changer-for-the-environment/

10   As an aside, all these considerations apply to strategic defense against nuclear attacks as well. They plagued the Reagan-era Strategic Defense Initiative and will plague President Donald Trump's Golden Dome program as well.

of ships. We may well find out the hard way someday how effective our defenses are—or are not. We can test and simulate, of course, and we do, but the Navy has, in my experience, resisted extensive high-fidelity testing, and the simulations and models in use are only as good as the data that goes into them. Resolving the question of surface ship survivability in a highly contested environment has to be central to future warship design and operational concepts, but that's not where we should start. We should start with the question of what missions need to be performed in the sea surface domain and what is the best way to perform them.

Those at all familiar with the way the Navy thinks about its missions in warfare have probably recognized by now that I'm not talking about the Navy's standard terminology of anti-submarine warfare (ASW), anti-surface warfare (ASuW), and air warfare (AW). Those concepts bring with them a certain way of approaching surface ship missions in terms of the targets one wants them to destroy as opposed to the missions one wants them to perform. There is a lot of overlap, but I'm trying to avoid starting with the assumption that we are talking about manned or unmanned surface combatants that need to kill certain types of targets and then optimizing the list of mission capabilities we want to get into a surface combatant hull to perform some mix of those tasks. If ships are expensive, vulnerable, and limited in speed, endurance, horizon, and mobility, then it may make sense to let platforms that don't have these characteristics do many of the jobs traditionally done by surface combatants. We only need to design operational concepts and corresponding ships for the situations where surface warships are the preferred tool. My provocative conclusion: This may be a fairly limited set of missions given what can potentially be accomplished more cost-effectively from the air, space, land, and undersea environments.

Let's start with the traditional Navy mission of destroying an adversary's surface navy. Naval warfare history is largely about fleet-versus-fleet engagements—Trafalgar, Jutland, Midway, and Leyte Gulf, for example. Until recently, there was no way to destroy an adversary's navy at sea except by creating a better navy, with some mix of superior quantity and quality. Other factors have mattered too—leadership, tactical brilliance, and training, to name a few. Nevertheless, technology has always driven the evolution of naval warfare in clear ways. Vessels propelled by oars that rammed opponents gave way to short-range cannon, wood and sail gave way to steel and steam, battleships with heavy guns and line-of-sight gunnery gave way to aircraft carriers with short-range aircraft carrying gravity bombs and torpedoes, and most recently, carrier-based aircraft with standoff anti-ship missiles are giving way to or at least in competition with ship-launched long-range anti-ship missiles. The constant in this evolution has been increasing engagement ranges away from the ship that carries the weapons, or in the case of the carriers and our construct, ships (transporters) that carry the launchers (tactical aircraft) that carry the projectiles or weapons (missiles). The reason ships were necessary to this construct is that until relatively recently, land-based systems lacked the range to deal with a distant enemy naval force. World War II in the Pacific was largely about acquiring islands from which fighters could provide air superiority and support to ground and naval forces and from which land-based bombers, with their superior range, could strike deeper and deeper into enemy territory. Because enemy fleets guarded those islands and posed a threat to our sea-based amphibious and naval forces, naval assets—carriers primarily—had to be employed to attack the enemy's fleets and suppress land-based aviation until it could be neutralized or overrun. When one compares the ranges of aircraft and missiles today (thousands of miles) to the ranges available then (hundreds of miles), and when one throws in the continuous surveillance now available from space, it's a totally different world. Maybe it's time to rethink some fundamental assumptions about how the surface of the sea is controlled.[11]

---

11   One way to think about this is that the separation between ships engaging
     each other has continuously increased over time. At some point as we
     separate the ships further from each other, we run out of ocean and into
     land. At that point it makes more sense for the ASuW mission to buy
     long range land based anti-ship weapons than it does to buy ships.

Costing over $10 billion each, aircraft carriers are expensive transporters of launchers and projectiles.[12] They are extremely high-priority targets for our adversaries, and during operations, they have to be continuously supported by refueling and rearming support vessels, which are also subject to attack. They are inherently limited in runway length and the size of the aircraft they carry, effectively limiting the range of the aircraft they transport. They are increasingly difficult to hide from a sophisticated enemy, especially when they are conducting flight operations, which restricts maneuver options because of the need to create wind over the deck.[13] Because of their small numbers and high operational value, other resources must be devoted to protecting them from a range of threats, including air, surface, and subsurface threats. At one time, through the end of the Cold War, a reasonable case could be made for the survivability of carriers. When I was Deputy Director of Defense Research and Engineering for Tactical Warfare Programs at the end of the Cold War, the Navy had convinced me it could conceal what was then called a Carrier Battle Group from detection, at least for an operationally useful period. This is increasingly difficult to assert convincingly. As threats have grown, the Navy has acknowledged that carriers cannot operate within operational range (a few hundred miles) of a sophisticated adversary's shorelines. As shore-based aircraft and weapons ranges extend, there comes a logical point when there is nowhere left for carriers and surface combatants to operate. At one time, carriers were the dominant sea surface control platform because of their aircrafts' lethal reach compared to guns. With long-range anti-ship missiles widely available now, that is no longer the case. In modern times, carriers have been used to provide continuing and cost-effective counter-air and strike operations over land but not against peer competitors. Against

---

12 This is just the cost to buy the carrier. One has to add the operating costs, the cost of the aircraft on the carrier, and several escort and supporting vessels to the total cost of a Carrier Strike Group. At any given point in time, only about one in three or four carriers is available to be deployed.

13 To launch fully combat-loaded aircraft, carriers generally have to steam into the wind to create wind velocity over the deck, effectively adding about thirty knots in relative velocity to the takeoff speed of launching aircraft.

lesser threats, this will be an enduring virtue and a good reason to keep the carriers now in existence, but it is not a strong argument to continue purchasing them at current rates indefinitely or to maintain them as the centerpiece of a sea control force.[14]

So, how should a future force be structured to defeat other countries' navies, support or enable power projection, assure the safe passage of commercial shipping, and provide sea control, especially far from the territorial US? Let's start with the mission to defeat an adversary's sea-based power-projection fleet. For the US, a goal here is to apply enough lethal counter-ship fire to defeat an invasion fleet operating against an ally or friend, such as Taiwan. As China increases its naval capacity to threaten Taiwan and assert its regional maritime claims, this becomes a pressing question. (This is almost the mirror image of the problem China has defeating US power projection from the sea, except China has a land-mass to operate from. It has chosen to solve this problem largely with land-based mobile anti-ship missiles along with other assets.) The first requirement is the ability to continuously detect, track, and monitor the adversary country's naval assets with targeting quality information. Ships are large and they have multiple signatures—visible, thermal, radar, and acoustic. They also produce wakes and emit signals to perform some military functions. An autonomous distributed network of surface-based sensors in a USV surveillance architecture concept was discussed earlier, as was space-based monitoring. Space would seem to be the most cost-effective approach to achieving this goal, if survivability or at least resilience could be assured (more about space survivability later). Unmanned long-endurance aircraft, possibly cued by other assets, are another possibility. In a robust force, some combination of space, airborne, and sea surface systems (all unmanned and autonomous) would be the most resilient approach and should be affordable. There are trade-offs among these options, but there is a high premium for redundancy and resilience. Let's

---

14   One can also make a case that carriers are useful after a sophisticated
       enemy's ability to conduct long-range engagements against
       carriers has been destroyed or suppressed by other means.

assume for now this problem is solved with a mix of reconstitutable, continuous, and long-endurance surveillance systems with secure communications to resilient and survivable command and control centers from which engagements can be planned and directed.

Our next goal is to provide the most cost-effective way to conduct those engagements. The target set for a peer competitor like China is nominally at most a few hundred vessels. If we are trying to halt a sea-based invasion of Taiwan, we would have a requirement to attack those targets within a few hours. We might have some period of warning, say, a few days, in which to mobilize and move assets closer to the target set. However, let's assume we don't have the luxury to move surface ships or submarines on station, which would require several days at least. We can expect the naval forces of concern to be reasonably well defended, with some long-range counter-air capability extending out to a few hundred miles and reasonably robust terminal defenses. Our projectile of choice is going to be anti-ship missiles, cruise certainly, and possibly ballistic and hypersonic. One could consider torpedoes or sea mines as well, and I wouldn't rule them out entirely, but a few considerations suggest that missiles make more sense against time-sensitive massed targets like an invasion fleet. (I'll cover submarine-launched torpedoes in the subsurface domain section.) Air-delivered torpedoes with ranges of over twenty miles could nominally circumvent terminal air defenses, but limitations of range, deconfliction problems with multiple targets, time on target, and delivery problems argue against them.

This brings us to a trade space for projectile or missile type and range, and for potential launcher range, capacity, and type. If land is available for the launch platform, that seems to be a preferred option, assuming mobile launchers, especially as with Taiwan, where we know the likely invasion target already. Land basing for anti-ship cruise, ballistic, and even hypersonic missiles provides opportunities for mobility, concealment, hardening, and deception, all of which are valuable against a sophisticated opponent. Land basing is favored if there is relatively local basing, probably out to something up to a thousand miles from anticipated engagement areas so missile range and therefore cost can be less. Shorter range reduces cost, but it also reduces flexibility, so the "right" answer isn't obvious. If suitable land isn't available, air-launched systems would be the next best option. Airborne launch platforms can be on station fairly quickly, are flexible in where they are applied, and with adequate standoff, the delivery platforms (launchers) should be survivable. They also offer some modest extended range by virtue of launch at altitude vice surface launch.

For a country like the United States, the best trade-off between weapon range and launcher range would seem to be to acquire enough weapon range to provide for survivability of the launching aircraft through stand-off, and no more. The launching aircraft is reusable, and the weapons are expendable, so lowering the cost of the weapon while tolerating some increased cost in the launcher makes sense. This gets more complicated when aircraft base survivability is factored in, which it definitely should be. (More on that in the air domain section.) The range and other features, such as capacity and cruise speed of the airborne launcher, are also dictated by the geography available to the acquiring nation and the amount of time needed to conduct engagements in the threat scenarios of interest. All this needs to be investigated through trade studies, but the first question we should try to answer objectively is whether or not it is now more cost-effective to use non-sea-surface systems to defeat an opponent's surface naval forces. I believe the answer is likely to be yes, and this has major implications.

Succinctly put, it leaves us with the problem of figuring out what other missions one needs surface combatants for. If carriers are not the preferred way of projecting power ashore against a peer competitor, and surface ships in general (carriers and large surface combatants) are not the preferred way of attacking an adversary's fleet of surface warships, what's left? What are the tasks or missions for which surface combatants (manned or unmanned) provide cost-effective military capability?

We have at least three missions to consider where surface ships have a potentially attractive role. We still want to provide cost-effective defenses for nonmilitary shipping, which prioritizes being in continuing proximity to the ships we want to protect as they move. We still want to provide continuous control of key maritime terrain where we might not have land access—choke points, for example. We may want to provide on-station quick reaction and responsive high-density strike or fires capability from the sea—in support of an amphibious assault or raids, for example. Even if we rely on airborne and space-based assets for surveillance and targeting, there is also some merit in having redundant sea-based sensing and continuous surveillance capability. What these missions have in common is the need for continuous presence and responsive fires. Surface ships on station or accompanying protected assets offer that important feature.

The first task future surface warships could perform relatively efficiently is the protection of other ships, or escort duty. The US is in the position of being separated by large oceans from the places where we would expect to be in a conflict with a great power. We are also dependent on global trade, primarily by sea. Our allies are on the other side of those oceans for the most part. Moving anything of substantial aggregate mass a long distance across oceans, like a large conventional force and all its support equipment and logistics, is much more efficiently done by sea. Transport by air is reasonably efficient for thousands of people, but not practical for deploying the many tens of thousands of tons of equipment associated with large-scale conventional forces and the logistics support for those forces. This is likely to still be true even for the largely unmanned ground forces described above. We also have allies that are crucially dependent on commercial shipping for essential imports, such as food and energy. The surface vessels that do these tasks are subject to attack on short notice and can best be defended by escort vessels that move with them and are designed to defend themselves, and the vessels they accompany, against air, surface, and subsurface threats. Commercial shipping is moving toward unmanned vessels as well, and one can imagine a convoy of sorts consisting of unmanned cargo and escort vessels.

Because the overhead associated with placing a small crew for command and control purposes on a commercial-sized ship or escort vessel with a reasonable sensor suite and munitions magazines is relatively small—a few percent of the capacity and mass of the vessel—the delta cost associated with keeping some humans aboard may be small enough to justify doing so, if only for unforeseen contingency purposes and to have someone "in possession" of the vessels.[15] One could envision some manning in peacetime and low-risk situations, but removing the humans for high-risk missions. Escort vessels will be expensive enough under any circumstances that high attrition would not be very acceptable. They would provide the last layer in a layered system of defenses—the terminal defense co-located with the defended assets in the traditional layered missile defense construct. Outer layers would be provided by other assets, to be discussed shortly, that would be responsible for broader sea and air control.

For the land warfare chapter, we assumed a force-on-force construct between a conventional current design force and a highly automated force. Let's start by assuming our problem is getting commercial shipping from the US to an ally in Asia successfully against an array of conventional threats: sea surface, undersea, and airborne. I don't expect China to restrict its military forces to within the second island chain. The concept is a tactical grouping of optionally unmanned autonomous vessels configured for the mission with a mix of weapons and sensors in a modular payload design that is loaded out consistent with threat expectations at the start of the mission. This tactical grouping can conduct surface warfare and anti-submarine warfare missions, but it is primarily sized for dealing with missile attacks whatever their origin. It would have to provide some degree of layered defenses, support extended-range launch on remote engagements, launch on warning, and provide terminal defense. The team of vessels would integrate onboard and offboard sensor data in a coordinated air picture and engagement strategy. Against large raids, all engagements would be automated.

---

15   One DARPA reviewer of the original report disagreed. His point was that zero-manning opened up previously unthinkable design ideas. One example he provided was removing threatening boarders by completely rolling the ship, something not likely to be considered in a manned ship design.

The weapons (projectiles) on the ships (transporter/launchers) are primarily going to be missiles. There will be substantial numbers of them and various types. There would be a standardized launcher system on the ships with various loadout options. Missile range and payloads and mix are tradeable and can be tactically adjusted, but the escort group would have to fight with whatever is loaded at the start of the mission/voyage. The ships involved have to carry reasonable payloads but not have too many eggs in a single basket, and they have to be vessels that can move with and somewhat more dynamically than the ships they are protecting. Sensors for the anti-missile mission can be a combination of offboard and onboard and passive and active. At least some of these escort vessels might include rotary-wing or possibly tilt-rotor surface and subsurface surveillance UASs. When all this is put together in a package, it looks roughly like something in a frigate- or destroyer-sized class of ship.

As indicated, the escort vessels provide the close-in defense of the protected shipping. Outer layers are provided by airborne assets that can target sea surface threats and, as much as possible, by land-based systems as well. The hardly original idea here is to destroy the archer, not the arrow. We discussed wider-area sea surveillance and land-based and airborne engagement systems earlier. If some threats—surface, subsurface, or airborne launcher platforms—through a combination of deception, stealth, and good luck, get within missile or torpedo range of the protected fleet, then the escort force also provides the final protection against those threats and their weapons.

An idea worth considering to augment this capability would be placing transportable mission modules on the commercial ships being escorted.[16] Both weapons and sensors could be carried and integrated into the defensive architecture. One could consider even substituting this capability for the dedicated military surface vessels described. My view is that

---

16   There's an analogy to the naval gun crews and weapons placed on merchant marine shipping in World War II. My father served as a Chief Petty Officer on one of these details in the Atlantic late in the war, after having served for most of the war on a destroyer in Arleigh Burke's "Little Beavers" squadron in the South Pacific.

some multi-role military capability is needed and can only be provided by dedicated vessels.

Our next surface combatant mission of interest is to provide continuous control of key maritime terrain where we might not have proximate land access—choke points, for example. In this situation, we are trying to prevent an adversary from transiting an operationally significant waterway. We may also be trying to protect the transit of friendly shipping. Here we are concerned with preventing a moving adversary force from transiting a fixed area at a specific time or with protecting friendly assets as they transit a fixed location where local threats exist. (For engineers, a Eulerian vice a Lagrangian frame of reference.) This is different in fundamental ways from protecting a moving set of friendly assets in the open ocean. Think straits of Hormuz or Malacca, for example, where no local power will allow the US to conduct operations from land, or an exit route from an Arctic bastion. In this situation, we have a need for continuously on-station, sea-based engagement capabilities of various types with relatively short weapon times of flight to the area of interest. The number of places where this situation would arise for the US may be relatively small. In most of the world, we have allies who might be willing to have US counter-sea systems deployed on their land, as long as those assets were used to defend the ally in question. It might be highly problematic to use those assets otherwise, however, so sea-based capabilities would be of value.

What we are trying to do here is to bring timely fires on a set of transiting surface targets. Those ships will have been detected and identified, and be tracked from space or by local sensors. The fires are primarily anti-ship missiles. The launchers for those fires can be surface ships with weapons-range standoff from the sea area of interest. That range would nominally be a few hundred miles at most, so that time of flight was reasonable for tactical responsiveness. The concept is essentially an arsenal ship. Fires could be cued and receive target tracking and updates primarily from space-based sensors, but also potentially from local autonomous sea-based, or airborne sensors. Synchronized raids could be planned and released using onboard or offboard command and control capability. US surface ship survivability would be through stealth, if possible. At one end of the spectrum, this concept is basically a barge with munitions. At the other end, it is a well-defended magazine with a mix of defensive and offensive weapons. The escort vessels discussed earlier, with a different projectile/missile loadout, could perform this mission. However, as I think about this operational mission, I'm not fully persuaded of the need for this capability from naval surface ships. Other concepts, such as on-station unmanned submersible weapon carriers or previously deployed smart mines, might offer advantages in cost and survivability, but with some disadvantages as well. If global surveillance is maintained through successful space control, or other cueing is available from whatever source with the lead time needed to put aircraft weapon carriers and their weapons in operational range, or to employ long-range ground-based fires, including hypersonic weapons, then the need for and advantages of surface combatants for this mission are lessened.

Finally, we may want to provide on-station, quick reaction, high-density strike capability from the sea. This could be a requirement in an amphibious assault or a raid, or in support of allied forces engaged on land. This situation implies a requirement for large amounts of relatively close on-station fires, over the horizon, but within, say, twenty to one hundred nautical miles. Fires would be both preplanned and responsive, as needed. The timing of those fires, within some operational window, is primarily our choice, although external factors, such as weather and enemy movement or synchronization with friendly forces, could easily affect the specifics. In addition to fires or strike, this mission would also bring a requirement for local air defense against anti-ship missiles. The construct described for commercial shipping escort missions would also apply here, but with the addition or substitution of modest-range offensive fires. One can also envision arsenal ships with a mix of fires, sensor platforms, and C2 platforms in some combination. The building blocks could be common with the escort vessels and those in the choke-point discussion (commonality improving the argument for that capability), but again with a different weapon loadout or mix of loadouts in each case—predominantly air defense for the escort mission, anti-ship for the choke point mission, and land-target strike for the fire support mission.

BUILDING BLOCKS

For control of the sea surface domain, we are left with a combination of a global C3BM system that integrates space, airborne, and sea surface unmanned sensors; long-range shore-based strike aircraft with medium-range anti-ship missiles; land-based anti-ship missiles; and surface combatants. I'll discuss the surface combatants here.

The basic surface combatant building block is a vessel with a suite of sensors and with the capacity and flexibility to carry a variety of weapons for either defense or strike missions. The vessel displacement would be in the nominal range of a mid-sized surface combatant—several thousand tons.[17] The vessel could be optionally manned, given that the overhead associated with a small crew for command and control would be modest. For a given multi-ship mission, most vessels involved would not have to be manned. We should be reluctant to rely on remote offboard C2 when operating against a sophisticated opponent with the capacity to mount a mix of EW, cyber, and kinetic attacks on communication systems and nodes, but we should be able to achieve secure local C2 of an operational formation of several vessels. Stealth is highly desirable, which implies low passive signatures in all wavelengths, a preference for passive sensors, and low probability of detection active sensors. Getting the balance of features right in this class of vessels will be a major challenge, just as it is in any surface combatant set of design trades. Those features include stealth, protection against the range of threats, the sensor suite, countermeasures of

---

17 On December 22, 2025, President Trump announced the intent to build a new class of crewed battleships, the Trump class. I have no idea how this decision was reached or of any analysis that would justify it. The last US battleships were built during World War II and retired in 1995. They weighed almost fifty thousand tons, roughly half the size of a modern aircraft carrier and several times the size of a modern destroyer. If built, the Trump class would be another high-value target for our adversaries.

various types, hardening, hosting of assets such as UAVs, damage control features, reliability, sustainment costs, and so on. This mix isn't much different from current design trades for a multi-role surface combatant like the recently canceled FFG(X) Constellation-class frigate, but the emphasis on survivability would be much higher than in current designs, and there would be design challenges associated with autonomous unmanned operation. Networked and potentially autonomous operational capabilities at the level of a multi-ship formation would be important for several functions, such as strike planning, air defense, and maneuvering.

In this concept, the unmanned surface ship serves as transporter and launcher. The projectiles are predominantly the missile systems for strike or defense carried by the USV. These would be evolutionary extensions of current missile systems and would potentially include hypersonic weapons.

COMPLEXITIES

Cross-Domain Considerations: Surface forces can deliver cost-effective multi-domain effects, but they also have critical multi-domain dependencies. Surface forces can provide support and effects to anti-submarine warfare and anti-air warfare, space surveillance and engagement, missile defense of ground targets, and naval shore bombardment or strike. However, because of the limited horizon of their organic sensors and the need to stand off from anti-ship threats, surface forces will be highly dependent on space assets and airborne assets for situation awareness and targeting for effective long-range engagements, in all domains. Surface forces will need to operate far from higher-echelon C2 and will be dependent on reliable long-haul communications. Surface forces are also likely to be part of a multi-domain force that conducts integrated space, air, surface, and subsurface operations rand support to operations ashore as an integrated whole. The concept provides for the option to have some humans for command supervision on some or all of the surface warships, but even with that feature, the need for reliable multi-domain-enabled communications to support operations is an absolute requirement in every mission.

Logistics: Warships stay at sea and operate for days, weeks, and even months without a return to port for maintenance and resupply. Nuclear propulsion aside, Navy vessels on station for long periods or conducting operations need to be refueled and resupplied at sea. It would be beneficial if munitions could be resupplied at sea, but today they are generally not, except for palletizable short-range munitions and some of the weapons delivered by aircraft from aircraft carriers.[18] If we take humans off the ships, some consumables and services are no longer necessary, but automated refueling and, if possible, rearming would still be required for the foreseeable future. In addition to resupply, maintenance needs for the many complex systems on a warship have been a major driver of crew size and composition. For the concept I've described to work, the design must be highly reliable, for the vessels themselves and for the systems they carry. These requirements are achievable and affordable; we are seeing similar requirements being met by other systems. However, high reliability and at-sea automated logistics support will be necessary unless we limit the envisioned warships to short-duration missions. It isn't hard to imagine automated refueling, which has been demonstrated.[19] As noted earlier, automated rearming for larger munitions has been problematic, but this is worth another attempt, particularly in the context of new designs that are not wedded to current standard Vertical Launch System (VLS) con-

---

18   The value of at-sea VLS reload has been clear for many years, but as noted earlier, the Navy has not been able to achieve this goal, although some progress is being made. Standardizing across ship classes on VLS has been of value for decades, but it has constrained weapons design and never been compatible with at-sea rearming. "USS *Higgins* Completes Expeditionary Missile Reload Simulation at Sea," *Seapower*, August 25, 2025, https://seapowermagazine.org/uss-higgins-completes-expeditionary-missile-reload-simulation-at-sea/.

19   "Automated Fueling-at-Sea Test Completed for Unmanned Surface Vehicle Program," DARPA, December 19, 2024, https://www.darpa.mil/news/2024/nomars-success.

figurations. Recent experiments have demonstrated at sea rearming of the VLS, however.[20]

Target Identification/Collateral Damage Avoidance: The sea surface uses of anti-air, anti-ship, anti-submarine, and land attack are all going to have stringent requirements for accurate target identification and avoidance of false positive and false negative errors. For longer-range engagements, surface ships will rely on target identification provided by offboard sensors. In many cases, surface ships will be receiving engagement orders and the actual engagement decision will be made at a higher-echelon battle command node. In some cases, data may be merged or fused on a ship to make a target classification decision on the warship, but I expect this to be the exception. For defense against large missile raids, the rules of engagement may be preset at a higher echelon, with individual ship and even flotilla-level decisions made locally. Onboard sensing dependency will occur in some self-defense situations against air and small-boat threats, especially in times of tension short of hostilities. In those cases, the actual decision may be made autonomously on the warship, but it is more likely to be made with human supervision or direct control of any engagements, either remotely or locally.

Design Requirements and Considerations

Anti-Tamper: The potential for a sophisticated opponent to acquire physical access to a warship has to be considered. Ships are lost at sea for various reasons, in peace and in war. While we would try to deny an adversary access to any vessel that had foundered or over which we had lost control, this can't be guaranteed. Critical functions will have to be designed with state-of-the-art anti-tamper. In addition to protection from physical and cyber hands-on reverse engineering, remote cyberattacks must be hardened against with high confidence. An obvious potential vulnerability

---

20   Dzirhan Mahadzir, "Navy Refines Rearming at Sea in East Coast Experiment," *US Naval Institute News*, July 23, 2025, https://news.usni.org/2025/07/23/navy-refines-rearming-at-sea-in-east-coast-experiment.

for autonomous warships is that control could be lost to a hostile actor. To some extent, this risk exists today for manned vessels also. Provisions for physically destructive anti-tamper and/or scuttling (either autonomously or on command) have their own risks but could also be included in the design.

Arctic and Antarctic Operations: I don't foresee significant new design requirements coming from the need to operate in these environments. The Arctic, especially, will be contested and there will be increased commercial traffic and activity. As a result, the US will need to protect assets in this area and attack adversary assets there, so warships will have to be designed to operate in arctic environments. Some design requirements will flow from the unique environmental features and threat possibilities in these areas. Icebreaker capability may be required and could be provided by manned or unmanned vessels.[21]

Auxiliaries: The Navy ship inventory currently includes a large number of specialized auxiliary vessels for various purposes, including replenishment and refueling, intelligence collection, ocean research, tugs, tenders, rescue and salvage, and command and control ships. The suite of auxiliaries needed to support the envisioned concept, in peacetime and in conflict, would have to be reconsidered completely. The survivability of these ships is low against future threats, so more resiliency would have to be designed in for future auxiliaries to be compatible with the operations of the new warships.

Civilian Engagement and Interaction: One attribute of current manned platforms is their ability to "show the flag" through foreign presence and port calls. Combatant Command leadership values this capability, as do diplomats. There are also several actions manned vessels can take, such as assisting vessels in distress, counter-narcotics and counter-piracy oper-

---

21   There has been long-running, and until recently unsuccessful, attempts to get the US Navy or Coast Guard to fund a small number of icebreakers. The Trump administration has recently announced an agreement with Finland to purchase eleven icebreakers. "Trump Announces Deal with Finland to Purchase Icebreakers," *The Hill*, October 9, 2025, https://thehill.com/policy/defense/5547900-trump-finland-icebreakers-deal/.

ations, and blockade or sanction enforcement, for example, that bring naval vessels in contact with civilians at sea or on land. These human-centric missions will not be automated anytime soon. For that reason, there will be a need for manned naval vessels in the force that can conduct these types of operations. I'd place them in the rough design category of the Coast Guard's Heritage-class Offshore Patrol Cutter.

Collision Avoidance: This is a basic requirement. All marine vessels are required to follow international and national collision avoidance rules called COLREGs. At this time, there has been enough experience with unmanned supervised and autonomous vessels that I don't foresee this as being a major design obstacle—if not already, then in the near future.

Confidence in Automated Behaviors: For surface ships and surface ship formations, the confidence in automated behaviors has to be extremely high and high across a lot of critical functions. Ships are big, and putting one in the wrong place at the wrong time has major consequences, as we have seen in relatively recent US Navy history.[22] (Although, arguably, a well-designed autonomous system might have done better than the crews involved in the collisions at sea over the last few years.) At this time, driven by economics, commercial technology seems to support unmanned ground vehicle navigation, at least on developed road systems, but the same can't be said for surface ships. Taking the cost of a driver out of a taxi or a truck has a more significant impact on vehicle life cycle total cost than removing the cost of maintaining a helmsperson continuously on a capital ship. Nevertheless, there has been a lot of international development in autonomous vessels. It isn't that the ground vehicle problem is easier, in some ways it's harder; the ground environment is more nonuniform. However, the steps a ground vehicle takes to avoid a serious accident are simpler and faster than those for a surface ship, and the consequences of errors are not on the same scale. One notable feature of capital ships, unlike ground vehicles, is that they have a hard time stopping and turn relatively slowly. For automated behaviors associated with tactical

---

22 There were two noted collisions involving US Navy surface ships in 2017 and another one in 2025.

decisions about engagements, the potential to engage a commercial airliner or a noncombatant vessel also carries heavy consequences. I would argue, however, that in peacetime, the absence of fear by those aboard for their own safety and the safety of the crew might substantially reduce the risk of an unwarranted engagement. The decision standards for wartime scenarios would be substantially different and tailored to the specific operational and tactical situation. In any event, we can attain and verify acceptable levels of confidence in these decisions, especially in a hostile environment where most threats are distinctive and provide distinguishable unique signatures. Tactical reality in the form of the time window to order defensive engagements against large attacks forces us to automate those operational decisions in any event.

Construction (Seabees): There will still be a need for forward and expeditionary shore installation construction to support the fleet. The degree to which this is needed, and the specific capabilities, would be dependent on the theater and types of operations planned. This support function would largely utilize state-of-the-art commercial construction technology.

Countermeasures: Countermeasures designed to defeat incoming attacks will have to earn their way onto the platforms in the concept based on their cost-effectiveness. Decoys like Nulka and EW systems like the SLQ-32 are good examples. These can be very high-payoff investments, but they also carry some degree of risk associated with unknown threat responses. Automated processes and streamlined, even real-time, software upgrades should be the norm in the future.

Cyber and EW: Electronic warfare capabilities of various types would be integrated into the warships in the concept. The radio frequency design features of the system would need to be integrated to balance and optimize several goals. I can envision a radio-silent mode, which, combined with other signature reduction, would make targeting more problematic for the adversary. Once concealment has been broken, the need for effective radio frequency support to close-in surveillance and engagement takes priority and active countermeasures would be enabled. AI technologies for smart radar communication systems and EW integration and machine learning are being developed now and would be included, along with the capacity for near-continuous, if not real-time, software updates. Cybersecurity would be a necessity. Active use of cyber to defeat attacking systems or against threat defenses is also a possibility that should be pursued and included in the concept.

Damage Control and Resiliency: One of the main reasons to have crews on warships has been their utility in damage control. Over time, the Navy has moved to more automated systems, but not entirely. The very flexible and adaptive ability of human beings in a complex, damage-induced crisis won't be replaced by automated capabilities anytime soon. The design in this area will have to accept that fact and do as well as is reasonably cost-effective to install damage limitation and reaction systems. Requirements for the resiliency of onboard weapon systems and the hull, mechanical, and electrical design to attack provide a broad trade space all the way from "ignore it" to design for continued operation despite multiple hits. The right balance is somewhere in between, of course, but the trade space looks very different when there are no people or a few people in a limited role to be protected, as well as the vessel itself to consider. Absorbing at least one hit from most classes of weapons and not sinking seems like an entry-level goal. Next on the hierarchy would be to retain the ability to maneuver and self-evacuate. Above that would be retaining degrees of mission performance. The Navy has plenty of experience designing for these types of requirements.

Deception: As stated earlier, it's increasingly difficult to hide surface vessels from the array of sensors that can be employed to detect, identify, and track them. It is also extremely difficult to win the cost-exchange ratio battle against the range and density of missile threats that can be employed against multibillion-dollar surface vessels once they are targeted. Deception, in the form of decoys meant to emulate vessels, could be very cost-effective. They would certainly be unmanned vessels and would need to emulate or anti-simulate real warships. This is relatively easy to do, but still not cheap even if only passive signatures are emulated. Emulating active signatures, for low probability of detection or low probability of intercept radio frequency emitters, increases cost but also increases deception effectiveness. Beyond the use of deception to emulate individual targets in a formation, actual operations could be simulated as well. The US has not embraced the widespread use of deception in general for a very long time, presumably because it diverts resources from "real" capability. Against a threat with a formidable lethal capability, the use of deception becomes much more cost-effective and should be considered, including for surface vessels.[23]

Directed-Energy Weapons: I've been watching directed-energy weapons, primarily lasers, be five years away from military applications for fifty years. Power levels, size, weight, and efficiencies have all improved over time. Warships provide a good environment for the volume, mass, input power, stability, and heat dissipation needed for high-energy lasers. At this point, however, I'm still agnostic about the contribution they will make. Lasers, and the optical fire control sensors that aim them, are limited by atmospheric conditions like fog and rain—both common at sea. They require nominally a few seconds of dwell time on an incoming missile to defeat it, limiting how many engagements they can conduct during the time window of a large, coordinated attack. They may not provide actionable feedback on their effectiveness, making the decision to move to another target high risk. The threat has a vote, and can use combinations of hardening, maneuver, and tactics to limit the effectiveness of lasers. Microwave weapons have similar problems. I'm open to the possibility of including directed-energy weapons in future warships, but they will have to earn their way by demonstrating their lethality, utility, and resilience first.

Electromagnetic Weapons Launch: This is another technology that has been worked on for decades and for which naval applications should be attractive because of the volume, mass, and power that ships can provide

---

23   We are seeing heavy use of deception in Ukraine by both sides, especially for ground forces. "Depend upon it, sir, when a man knows he is to be hanged in a fortnight, it concentrates his mind wonderfully." Samuel Johnson as quoted in James Boswell's *The Life of Samuel Johnson*, LL. D, vol. 3.

or accommodate.[24] The concepts I've described could include this technology for long-range strike and for self-defense, but it is going to have to demonstrate its relative cost-effectiveness over other options first. At this point, I'm still skeptical of the cost-effectiveness of this technology.

Energy: Future military surface vessels as envisioned here will employ the most efficient energy sources available for propulsion and onboard systems when they are fielded. At this point, I would anticipate the infrastructure to support refueling, including, in all likelihood, at-sea refueling, which will have to be included in the concept. Small unmanned surface vessels can utilize nontraditional energy sources, such as wind, solar, and wave riding, for long and very long endurance missions, but with penalties in speed and maneuverability.

Fast Attack Craft and Fast Inshore Attack Craft: These are primarily a coastal threat to large warships and commercial traffic operating in confined areas or near shore in general.[25] Swarming small vessels like these provide a potential way to attack capital ships cost-effectively. Any vessels in the concept that would operate where they could be subject to this sort of attack would have to be armed with the means to defeat them. A modular approach to providing that capability when needed seems reasonable. The degree of human control, on board or remote, of such a defensive system would need to be flexible depending on the circumstances. I do not envision a swarming small craft option for the US because of the problems associated with delivery of those craft to where they are useful and because they would be essentially all expended when used. Delivery is necessary because of their limited ranges. Also, either they wouldn't survive or their trackable return to the mother ship would likely be fatal to that vessel.

---

24   Electromagnetic aircraft launch is on the new Ford-
class carrier, which has now entered service.

25   Interestingly, Thomas Jefferson once sponsored a similar concept for
US coastal defense—swarming small gunboats with a single cannon
operated by part-time militias. It was tried in the War of 1812 but didn't
work out very well; none of the gunboat commanders wanted to go first
against a British ship of the line with a full broadside available.

Intelligence Integration: The C3BM architecture would need to flexibly integrate threat and other relevant intelligence for a wide range of sources, including air and space, in particular. Organic sensors would only provide a fraction of the information associated with attacks or targets for strike. Onboard-generated information might be sent to a shore-based node for inclusion or integrated with information sent from shore as it arrived. In general, however, operational and tactical intelligence integration would take place under human supervision where it made the most sense to integrate the data—generally not on the warships, even the manned warships if they were part of the formation. For defensive purposes, the necessary intelligence, processed to meet onboard and group tactical automated decision needs, would be pushed to the warships, along with governing rules of engagement. Until engagement planning needed to occur, the ship or formation would only need limited information to carry out instructions. Once engagement planning was necessary, early engagement (launch on remote) instructions would be sent.[26] At some point, as threats closed in time and distance, organic sensors and organic battle management would begin to control engagement decisions. For strike missions, intelligence would generally be integrated in nodes ashore and strike orders would be passed to the warship formation. Throughout the architecture, edge processing of raw intelligence sensing should be emphasized to limit data flow (bandwidth) requirements.

Interoperability: Several years ago, the then-Chief of Naval Operations articulated a concept of an integrated several hundred ship collective naval capability with our allies. That's a sound concept, but achieving it with the concept I've described will take unprecedented degrees of interoperability. At some point, our partners might adapt similar operational concepts to the one I've described, but for the foreseeable future, one can expect a mix of more traditional allied vessels and the types of autonomous vessels

---

26 Launch on remote refers to launch decisions from host weapon platforms based on threat tracking data from offboard remote sensors.

I've described, possibly together with manned US vessels.[27] We will need to be technically interoperable with our partners to be effective. The US is constantly conducting naval exercises with friends and allies around the world. We would want that practice to continue for several reasons. I expect interoperability would be achieved incrementally. The first step would be some integration of individual or small numbers of allied vessels with the warships I've described. As now, we'd need means of intelligence sharing and integration of the allied vessels into US formations, or at least the ability to communicate, share data, and operate cooperatively. Past efforts to achieve this sort of integration have had mixed success at best, but that is as much a matter of prioritization and command attention as it is technology.

Humans in the Concept: The critical humans in the concept are the ones in the command organization, either ashore or embedded on a ship in the formation, overseeing the operations of the force and providing executive control. I did provide for a small C3BM team on a subset, or even on each, of the warships as an option because the overhead cost was low and one could envision cases, especially in peacetime or tense situations short of conflict, where having some humans on board would be of high value. There would also be humans involved in remotely overseeing the continuous operation and the nearly continuous updating of the software associated with operating the combat systems on the ships. Embarked humans would need to have all necessary life support on the ship, but perhaps not for the long durations supported by current designs

Hypersonic Weapons: They would be included based on cost-effectiveness. The ships could be designed to host ballistic hypersonic boost-glide or air-breathing hypersonic cruise weapons. These weapons may not be cost-effective in large numbers for strike missions against the target sets of interest to US commanders. However, they do have better penetration capability against current defenses and shorter times of flight than transonic cruise missiles. One has to weigh the costs and trade-offs between

---

27 My assumption is that the US will lead the operational implementation
   of lethal autonomy. That may well not be a valid assumption.

these and other options for the target sets of interest. Intuitively, some of these weapons, delivered by some means, make sense in the US inventory. I'm not persuaded at this point that those numbers are large, or that surface ships are a preferred host for them.

Legal Constraints: There are several legal constraints that apply to the overall concept and some that apply to the warships I've described. There is nothing unique in the sea surface domain that would be especially problematic. COLREGs compliance was described elsewhere. Humanitarian and law of armed conflict provisions would apply to the use of force as in other domains. There are unique maritime legal considerations, navigation and position reporting requirements, salvage law, waste disposal, and other environmental regulations, such as marine mammal protection, fisheries, customs, quarantine, and piracy, but nothing that precludes the concept that I think is unworkable or that the Navy doesn't address routinely.

Loss of Contact/Control by Echelon: This is an area where reliability and resilience are particularly important and where offboard sensor-based situational awareness is an absolute necessity. The ability to track a vessel's location should be multiple-redundant and failsafe. The ability to determine status and take basic control over maneuvering should be almost the same level. Default behaviors implemented on board in the event of loss of external communications or partial organic system failure should be tailored to the situation and monitored and updated continually.

Marines: Large-scale amphibious operations aside, Marines are often embarked on warships to provide a contingency capability and provide physical security against some threats. For the core escort and strike missions the concept was designed for, they aren't necessary, but the flexibility to have a support module to embark a squad or even somewhat larger unit of Marines on board would be an option that could be designed in.

Operational Planning and Rehearsal: This would take place at the C3BM nodes, likely ashore. Rehearsal would be virtual, of course. Prior to a deployment or operational mission planning, war-gaming, alternative course of action analysis, and virtual rehearsals could all occur in a synthetic environment, but it would certainly make sense to include the live onboard systems "in the loop." During operations of any type, planning would be updated continuously as the tactical situation changed.

Organic Aviation: An option should be provided to embark organic aviation. I don't believe every vessel should have organic aviation for every deployment or mission, but it would be a valuable asset to have in a few circumstances. Organic or at least warship-compatible unmanned aircraft could extend local high-resolution surveillance for various missions, provide for evacuation and support to vessels with humans embarked, and provide an extended-range communication link. In some situations, they could also be armed to counter relevant threats. An advanced design roughly in the category of the Fire Scout MQ-8C model seems about right. A tilt-rotor or other advanced vertical takeoff option would be possible as well. Such a system would have to be reliable enough to operate without continuous onboard support, and it would have to be capable of being refueled or recharged on board autonomously.

Physical Security: Some security would be provided by the basic design, which should defeat unauthorized entry for low-end threats. Nonlethal effects could be used to provide additional deterrence. Anti-piracy techniques might be relevant in some cases. I wouldn't rule out remotely activated and operated, or even in some circumstances, autonomous lethal defense mechanisms. In port, more traditional security can be provided as needed. Underway, the suite of capabilities described here could be activated, with degrees of severity depending on the threat and the situation.

Port Facility Support and Pilots—Overseas Basing: In a world in which future autonomous commercial shipping comes and goes, I don't see a fundamental obstacle to overseas basing. The vessels would have the provision to be optionally manned and could take a pilot aboard if required to enter a port. More likely, as I expect for future commercial vessels, the pilot function will be performed remotely without the need to board the vessel. The same could be done for the warships in the concept. Anytime one of the envisioned vessels entered a foreign port, it could be met by humans (Navy or contracted) for in-processing, assessment, arrangement of needed maintenance, refueling and rearming, and to provide physical security.

Reliability: This may well be the biggest obstacle to realizing the described concept. Redundancy can help, but some shipboard systems cannot be redundant because of their size and/or cost. Many critical supporting systems are made of components or assemblies that can be changed out at sea from a spares inventory carried aboard, but not as a practical matter on an autonomous ship. I do not envision a general-purpose highly flexible electrical, mechanical, or hydraulic "repairperson robot" anytime soon. The consequence of an unrepairable casualty at sea can be catastrophic, causing a return to port or the need for a salvage operation with offboard assets. Of course, this is somewhat true for manned ships also, but spares for high-failure-rate components can be stored aboard and installed at sea as needed on manned ships. That won't be the case here. Humans are very good at replacing complex parts—this function isn't going to be fully automated in a general way for some time, so high reliability is a necessity. Fortunately, a lot of progress is being made in this area, especially for commercial electronics and other components where the market rewards high reliability. Unfortunately, a lot of shipboard parts don't currently meet this description. On the plus side, one thing the current experimental unmanned vessel program is achieving is the maturation of technology in this area. The DARPA NOMARS (No Manning Required Ship) program, for example, aims to demonstrate a full year at sea on a medium-sized vessel.[28]

Replenishment at Sea: Refueling at sea is likely to be a necessity and would have to be automated.[29] With current technology, an autonomous or minimally manned vessel of the size envisioned could have an unrefueled range of several thousand miles. That isn't enough. This function

---

28 David Szondy, "DARPA Christens Dagger-like, Crewless Ocean Warship," *New Atlas*, August 13, 2025, https://newatlas.com/military/darpa-defiant-crewless-ocean-warship/.

29 It depends on the unrefueled design range. Ships in the class we are talking about probably couldn't conduct a subset of needed missions (Pacific transit plus operational maneuvering as needed, for example) without refueling.

could be automated, but a lot of work will have to be done to demonstrate it can be done effectively and consistently in a reasonable range of sea states.[30] DARPA and the Navy have already demonstrated a limited capability to conduct at-sea refueling of an unmanned vessel.[31]

Requirements Creep: Conscious attention should be placed on avoiding cost imposing marginal or low return-on-investment requirements. The natural tendency in all domains is to add more requirements to the design until the concept crashes from its own weight. A lot of the items on this list are good examples of potential requirements creep. The whole point of the concept is improved cost-exchange ratios over current systems.

Responsive Threats: The threat spectrum isn't significantly different than it would be for new manned warships. The potential Achilles' heel of a cyber or EW attack on an unmanned ship would be very attractive to an adversary. Just isolating the ship from its C3BM system would have a crippling effect if it prevented integrated multi-vessel extended-range operations, for either air and missile defense or strike applications. We can expect the threat missiles to achieve longer ranges, employ countermeasures like decoys (in the case of ballistic threats) and self-protection jammers, adapt maneuvering and collaborative tactics, and use attack structures tailored to the defending force. When dealing with a thinking, competent, and well-resourced adversary, we will need to be at least a step or two ahead in analyzing potential responses and either designing for them or making provisions for growth paths to respond when they materialize.

Salvage/Recovery: Bad things, short of outright sinking, can happen to ships, necessitating salvage or recovery. The ships envisioned here will not be expendable, and there will be situations even in peacetime where

---

30　The Navy Sea Hunter USV is reported to have a range of ten thousand nautical miles. The Next Generation Large Unmanned Surface Vessel, L-USV, concept is intended to have a range of at least thirty-five hundred nautical miles.

31　"Automated Fueling-at-Sea Test Completed for Unmanned Surface Vehicle Program," DARPA, December 19, 2024, https://www.darpa.mil/news/2024/nomars-success.

salvage is required. One provision to consider is the idea of recovery of one autonomous warship by another. The ability of salvage and recovery teams to come aboard and deal with a range of possible equipment casualties should be provided for in the design. This function won't be fully automated anytime soon, although small autonomous vehicles could do some aspects of this function.

Single Points of Failure: This is a subset of the reliability requirement, but it specifically addresses nodes or functions essential to operations and forces consideration of backup or lesser-included capability alternatives (graceful degradation) and a means of reverting to them as necessary. The autonomous operating system for the vessel and its principal weapon systems and sensors, as well as the C3BM architecture, are all critical to mission success. This can't be done for every function in the so-called kill chain, but it should be part of the design process.

Stealth: There are different types of stealth for different threats. Defeating even a subset of the adversary's sensors can have value. It may not be possible to hide the presence of a vessel from some types of observation, space-based wake detection for example. However, other aspects of the system, the fact that it's a warship, or the ability of a threat seeker to detect the ship or to identify its most vulnerable area, might be defeated. Stealth, in conjunction with deception, can have symbiotic effects that merit its inclusion in the design. Operation in a passive mode was discussed earlier. Low probability of detection and low probability of intercept radio frequency emissions through various techniques can contribute significantly. Reduction in radar and optical signature or creating confusion about identity can have value as well. The Navy has worked on each of these approaches, but the overwhelming nature of the threat is pushing us toward the need for concealment in lieu of active defense as a much more significant contributor to survivability. USV design should reflect this reprioritization.

Training, Experimentation, and Testing: One virtue of unmanned sea surface concepts is that there is no need to keep ships at sea on a continuous rotational training cycle to maintain proficiency. Unlike humans, who tend to lose competency unless it's exercised frequently, computers don't forget. That doesn't mean there should be no exercises or live training, but the need can be significantly reduced, and much more can be done through virtual and constructive means. Allies can be involved in this as well to improve interoperability. Some traditional testing—shakedown deployments, sea trials for new construction and systems, and so on—would still be required.

Unattended Sea Surface Sensors: Unattended sea surface sensors are generally slow moving or drifting, but they can be proliferated economically, be expendable, and have high utility. They can be made very difficult to detect. The options include wave gliders, sail drone concepts, and drifting buoys. These systems can also be used by adversaries as part of their overall ISR suite, and in some cases (such as choke points or sea lanes), they can be weaponized as well and be very cost-effective mines. I can see both the US and its adversaries using these systems, possibly extensively. As their use increases, this will levy significant requirements on ISR systems in space, air, and sea surface domains and create a demand for cost-effective neutralization mechanisms.

Weapons Mix: For the surface warships in the concept, the design should accommodate flexibility in weapon loadout for the operational mission, using common interchangeable weapon modules where possible. The mix could include some combination of anti-air weapons, conventional and hypersonic strike, anti-ship missiles, anti-satellite weapons, remote delivery of unattended sensors, and possibly other weapons. It would be highly desirable for these modules to be replaceable at sea.

Weather Implications and Forecasting: I don't foresee any novel weather-related design requirements different from those for current systems. It might be possible to push for higher sea state capabilities and relax some of the current motion and stability-related requirements for human-operated warships in those cases where the vessel is operating unmanned and autonomously. I'm skeptical that this would have high payoff, but it should be evaluated. Regarding forecasting, the needs would be similar to those for current operations, with even more precise forecasting desirable. Resilient space-based systems would be needed to provide this support.

Other Needed Military Functions

Amphibious Operations:[32] Traditional amphibious assaults against prepared peer competitors are already problematic because of the vulnerability to precision weapons of the transporting surface ships and landing craft, and the relatively slow pace at which any amphibious landing can be conducted.[33] Smart, or even traditional, sea and land mines compound this problem. Just the process of loading, deploying, and navigating landing craft ashore can take hours, and the deploying ships must be within relatively short range of the landing sites, even if they can be held over the horizon from shore. The Marine Corps has tried to speed up this process with greater use of aerial delivery and faster landing craft, but the fundamental problems have not been overcome. The Marine Corps has also attempted to develop doctrine and concepts that increase the probability of successful amphibious assaults. Most recently, it is shedding some of its heavy and less transportable equipment, such as tanks and towed artillery.

Despite these efforts, it's still a major undertaking to get Marine units ashore and support them after a landing. If the supporting ships cannot survive or remain on station to support the assault force, it will likely be defeated, whether it is composed of primarily autonomous systems or more traditional ones. As a result, the best alternative for forced entry from the sea may be to have the capacity through air-delivered weapons, enabled by air- and space-based ISR and C3BM, to neutralize defenses so the assault force is weakly opposed or unopposed. "Going where they

---

32   This topic is worthy of its own chapter or an entire book. There is no more challenging type of military operation. Against a prepared, well-equipped, and trained foe, large-scale amphibious assaults pose enormous challenges.

33   The Taiwan invasion scenario confronts China with all these challenges and provides a strong disincentive, assuming Taiwan has acquired and is prepared to use a cost-effective mix of sensors and weapons.

ain't" is another theoretical option, but difficult if not impossible to achieve, given the ubiquity of sensor systems to detect an assault force and the availability of long-range anti-ship precision weapons with target detection and selection capabilities. Suppressing the ground mobile long-range precision missiles that would be used to target amphibious shipping would be a daunting task, but it isn't unthinkable. It would be desirable or even necessary to establish dominance in the air and in space prior to attempting an amphibious assault. Once that is accomplished, one can envision a highly autonomous suite of systems, including those discussed in the land domain section, and compatible transport shipping as the basis for a forced-entry ground force. The transport shipping would be designed specifically to support the rapid movement of a UGS and UAS force ashore, as described in the land domain section. The best delivery means, because of their speed, would be aviation assets as opposed to landing craft. The organic UASs in the land domain concept would be the first wave ashore; they would eliminate short- and intermediate-range threats and watch over the subsequent air delivery of their hosting UGVs ashore. Once the full force was ashore, it would function identically to the land domain concept. Implementing this concept would require almost a complete redesign and recapitalization of current amphibious assault platforms.

Anti-Submarine Warfare: Sea surface assets can be vulnerable to attack by submarines using torpedoes or standoff cruise missiles and ballistic missiles. Surface vessels and their organic aviation assets also have a role in ASW. As I'll indicate in the next section, the ASW role is moving toward air- and space-based, as well as undersea systems. Nevertheless, surface warships can carry both sensors and weapons that can be useful in ASW operations. Surface vessels must be able to defend themselves and other assets from missile and torpedo attack, whatever the source. Defense against torpedo attack is best achieved by attacking the launching platform, submarine or otherwise, but in addition to decoys, there is some anti-torpedo capability fielded and in development.[34]

Blockades and Embargo Enforcement: If the missions involved stop and search, some manning and even a Marine or other armed boarding-party module would be desirable. The offshore cutter-class design would be suitable for this function as well. If the situation was hostile and the blockade was lethal, the calculation would be different. Unmanned systems with offboard C2 could be effective. In that case, however, space and airborne capabilities might be a preferred choice.

Counter-Piracy and Counter-Narcotics: These are law enforcement missions and should not be major design drivers for naval warships intended to prevail against peer competitors. The concepts defined above could contribute to these missions and would be expected to do so.

Disaster Relief/Humanitarian Assistance: The US military does these missions effectively with some specialized assets, like hospital ships and small-deck carriers, as an inherent capability. My view would be that we shouldn't buy military assets specifically for these missions, but we should certainly use the military assets in the inventory where they can be effective. I would expect the US to have relevant assets for these missions for the foreseeable future, but perhaps not in the current density.

Freedom of Navigation Operations: The autonomous vessels in the concept could certainly do these missions, but if unmanned, they might be tempting targets. This is another, but not necessarily compelling, reason for optional manning, even in small numbers.

---

34 Active anti-torpedo self-defense systems in the US have been and are being attempted, but with limited success so far. US Navy fielding is at best a few years away. Carter Johnston, "US Navy Sets Sights on Fleet-Wide Anti-Torpedo Weapon Rollout in Coming Years," *Naval News*, May 7, 2025, https://www.navalnews.com/naval-news/2025/07/u-s-navy-sets-sights-on-fleet-wide-anti-torpedo-weapon-rollout-in-coming-years/.The German Navy is also continuing its Sea Spider program, which may enter production in 2026. "Germany to Buy Anti-torpedo Torpedo in 2026, Leaked Document Shows, *Defense News*, Sept. 23, 2025, at: https://www.defensenews.com/global/europe/2025/09/23/germany-to-buy-anti-torpedo-torpedo-in-2026-leaked-document-shows/.

Mine and Countermine Warfare: Future surface domain concepts overall could include sea mine delivery options, but they would probably be better delivered covertly by undersea systems or possibly by air. (I'll discuss smart mines in the undersea section.) Countermine warfare is an issue for amphibious assault forces and when friendly shipping (military and commercial) has to transit areas where mines can be emplaced and be cost-effective—at choke points, for example. The undersea domain is an extremely difficult countermine environment. The potential for false alarms, the opportunities for concealment, increasing smart mine options, and the relative ease of providing decoy mines to delay and increase the cost of clearing are all problematic. The US Navy and others have pursued autonomous vehicle-based solutions, deployed from surface ships primarily, and airborne sensors for some time with at best mixed success and some failures. This effort will continue and can be increasingly unmanned and automated, but I will be surprised if it ever achieves high-confidence rapid mine-clearing capabilities. Reliable mine field detection (as opposed to detection of individual mines), avoidance, and relatively slow clearing processes, all done autonomously, may be the best we can expect.

Naval Bombardment/Fire Support:[35] This is something to consider, but not likely to be a major design driver. The warships in the concept could provide this with missile systems, but with less volume of fire than gun systems for a similar-sized vessel. Gun systems aren't out of the question, either electromagnetic-launch or more conventional. They would have to earn their way into the concept through competitive cost-effectiveness. Against a peer competitor, the survivability of any surface vessel providing fire support to forces ashore is questionable, but in many operational scenarios, it could be cost-effective, particularly against threats where the anti-ship kill chain had been suppressed (see the amphibious

---

35   Early in my career, in Vietnam and later in Lebanon, the Navy was still
      using World War II vintage battleships to provide naval bombardment.

operations discussion above) or in something analogous to an "artillery raid" concept for ground forces.[36]

Regional Missile Defense: Aegis-equipped ships currently provide a contribution to regional missile defense of assets on shore. The ability to move these ships to a given theater, to the proximity of protected assets, or to an optimal point from which to conduct intercepts against a given threat, provides useful operational flexibility. The envisioned surface warships could perform this function, and to the degree which their presence can be concealed, they could make a threat attack planner's problem more difficult and higher risk. Their mobility would also improve their survivability once they start conducting intercepts.

Special Operations: Presumably some types of special operations conducted from the sea, such as raids and support to covert operations of various types, would still be needed. The surface warfare concept wasn't designed for those missions, and specialized surface or subsurface vessels for those purposes might still be needed. We won't be fully automating these complex functions anytime soon, and some of them require human-to-human interactions. Some subsets of special operations could be conducted with automated systems, however, such as covert delivery of support to insurgent groups or loitering covert weapons for surgical strikes on specific targets.

CONCLUSION

There has been a significant amount of maturation of unmanned surface vessel technology as far as basic vessel reliability, endurance, refueling at sea, and navigation since I wrote the original report for DARPA a few

---

36   An artillery raid moves weapons forward within range of enemy fire for use over a brief period, after which those assets are withdrawn before they can be engaged. With modern ISR, counterbattery, and tight C3BM, this is very hard to achieve. Something similar has been proposed for aircraft carriers—short excursions within the range of missile threats. Given how long carrier operations take and how detectible they are, this seems like a highly risky proposition to me.

years ago. There has been less progress in determining what role these vessels might play in the fleet, or even about the role of surface combatants and aircraft carriers more generally. As long as the thinking is constrained by the assumption that uncrewed vessels will be designed to augment or complement traditional crewed vessels, that may remain the case.

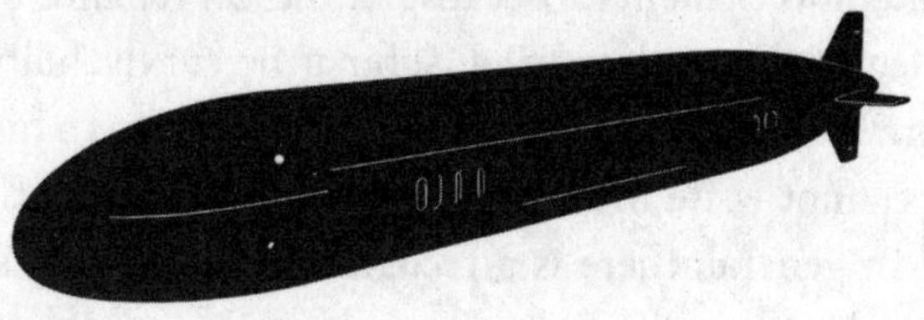

CHAPTER 4

WARFARE DOMAINS— SEA SUBSURFACE

The subsurface domain was identified during the work on Competitive Strategies and the Third Offset Strategy as one of the few areas—maybe the only one—in which the US retained a substantial advantage over its near peer competitors, especially China. The US advantage is largely in the areas of stealth (acoustic quieting, primarily) and sensing. While submarines do carry some defensive countermeasures to defeat attacks, they are very limited. Submarines achieve their survivability largely through not being detected. Over the modern era, efforts to improve active and passive acoustic detection have received the most attention, with heroic efforts to improve submarine quieting and resilience against active sensors through sound absorption and other techniques. There are two problems with relying on stealth so heavily for survivability: First, it can be fragile—once it is lost, submarines become very vulnerable; and second, it limits the things a submarine can do operationally.

For years, major powers have worked to defeat submarine stealth. Submarines are fairly large metallic objects with a variety of signatures— acoustic being the one that has received the most attention since submarines came into existence.[1] Because of the US reliance on the submarine-based leg of the nuclear triad, submarine survivability has been of grave concern for several decades. Submarine warfare is a highly classified topic and I cannot write in much detail about the current state of ASW technology. In general, there is a recognition that hiding submarines is becoming harder as sensor technology, the ability to detect novel signatures, and the capacity to pull signatures in general out of the background noise all improve. Quantum sensing, artificial intelligence data analytics techniques, machine learning, and the proliferation of sensor systems at sea and elsewhere are working against the continuing efforts to improve the concealment of submarines. I can't comment here on the precise state of that competition, but time and technology do not seem to be on the side of the submarines.[2]

Like modern surface combatants, submarines are big, expensive, and slow. When submerged, their ability to sense beyond the immediate environment is very limited. A Virginia-class attack submarine is four hundred feet long, weighs about eight thousand tons, and has a crew of 135. Unlike surface combatants, submarines are relatively survivable, until they try to go fast, fire off weapons, or operate at or near the surface. Speed means creating much more noise in the water; firing weapons—either torpedoes or missiles—creates a large signature. Coming to periscope depth to collect information significantly increases detectability. Unlike surface combatants, submarines also have almost no capability to defend themselves once they are detected.

---

1   I've spent part of my career on the anti-submarine problem, going back to the Air Defense Initiative in the 1980s, which focused in large part on countering Soviet submarines that could launch nuclear-armed cruise missiles from off the shores of the United States.

2   My friend and one of the people I respect most in the national security arena, Dr. Bill Perry, wrote a book in 2020 advocating elimination of the intercontinental ballistic missile leg of the nuclear triad and more reliance on submarines. I rarely, if ever, have disagreed with Bill, but I think it would be very high risk to shift our nuclear deterrent to rely so heavily on the survivability of submarines. William J. Perry and Tom Z. Collina, *The Button: The New Nuclear Arms Race and Presidential Power from Truman to Trump* (Dallas: BenBella Books, 2020).

The threat to submarines, in addition to other submarines, includes fixed and deployable sensor systems that have been employed to geolocate submarines, particularly as they transit confined waters. Detecting a transit provides queuing for more localized sensors and for other submarines to track the detected submarines. It's one thing to search the open ocean for a submarine that may or may not be there. It's considerably easier to stay on top of a known submarine once it's been detected and geolocated. If one has the freedom to operate them, airborne ASW platforms, like the shore-based P-8 with deployable sonobuoys and other sensors, and the ship-based MH-60R helicopter with a number of sensors, provide a significant threat to current submarines. It's been a truism that someday submarines will become detectable, and their relative security will be fatally compromised. To most observers, that day has not yet come, and it always seems to be a few years away, at least.

However, as I look at emerging technologies, such as those mentioned above, I'm convinced we cannot count on the long-term viability of submarines as we currently build them. In 2016, I asked the then Chief of Naval Operations, a submariner, what his views on this were. I won't repeat his answer here, but it was not reassuring. If we take it as a working assumption that current large, crewed submarine designs and concepts will become vulnerable to detection and attack (at least when operating in waters of interest within a thousand miles of adversary coastlines) somewhere in the five to ten -year window of this book, where does that leave warfare in the future in the undersea domain, and is there a role for lethal autonomous undersea systems?

Considering the potential for unmanned undersea vessels opens up some significant design trade space but also brings some challenging operational problems. The Navy has been experimenting with undersea autonomy for some time. Until recently, this generally took the form of small unmanned undersea vessels (UUVs) that could be deployed from surface vessels, through submarine torpedo tubes, or from the decks of manned submarines.[3] Countermine operations and local surveillance and reconnaissance missions have been explored to varying degrees. For the last few years, the Navy has been evaluating and experimenting with new concepts for what it calls subsurface and seabed warfare (SSW), a phrase that captures the mix of manned and unmanned undersea vessels, fixed and mobile sensors, and seabed infrastructure for communications, power, and other uses. This is promising, but the potential exists to go much further. Most recently, larger experimental designs with longer-range surveillance or with mine and sensor delivery missions have been developed, and are entering serial production.[4] Internationally, there are a number of efforts now underway to develop and field large UUVs.[5] For these, and any unmanned undersea system, a major issue is command and control, simply because of the physics associated with undersea signal transmission. This can be mitigated in various ways, but with associated penalties in detectability and mobility. Let's see what operational capabilities might be feasible in the not too distant future.

3   The Navy recently demonstrated the ability to recover a torpedo-compatible small UUV. Michael Fabey, "REMUS UUV Completes First Virginia-Class Torpedo Tube Recovery and Swimout," *Jane's Defence Weekly*, October 8, 2025, https://www.janes.com/osint-insights/defence-news/defence/remus-uuv-completes-first-virginia-class-torpedo-tube-recovery-and-swimout.

4   Most notably, the Orca XLUUV, which Boeing and Huntington Ingalls are now building five of for the Navy, and which the 2025 Navy budget procured at a rate of one per year. At this point, these are experimental systems, and the Navy is working to determine what payloads and operational missions would make sense for the platform, with mine delivery the firmest application. Ronald O'Rourke, *Navy Large Unmanned Surface and Undersea Vehicles*, R45757 (Washington, DC: Congressional Research Service, March 25, 2025),https://www.congress.gov/crs-product/R45757.

5   An example is Ukraine, perhaps the most highly motivated military innovator on the planet. "Secrets of Ukraine's Toloka Naval Drones: What the New Underwater Family Can Do," *Defense Express*, September 22, 2025,https://en.defenceua.com/weapon_and_tech/secrets_of_ukraines_toloka_naval_drones_what_the_new_underwater_family_can_do-15898.html.

OPERATIONAL CONCEPT

I'll assume we are primarily interested in defeating enemy submarines and surface ships using assets that operate in the undersea domain. Other lethal missions are possible, such as attacks on land-based targets and covert mine emplacement, but ASW and ASuW seem like the obvious principal motivation to consider lethal autonomous undersea platforms.[6]

If one accepts that reliable long-endurance UUVs are plausible, something that I think has been conclusively demonstrated in the last few years, and that some degree of lethal autonomy in the undersea realm is acceptable, then it opens up some interesting possibilities. The first mission we should consider in this domain is ASW—defeat of an enemy submarine force. Traditionally, submarines are thought to be the most effective anti-submarine platform and torpedoes the most effective anti-submarine weapon. (I'd argue that if they could be successfully built, wide-area airborne sensors or air-delivered sensors from fixed-wing and helicopter platforms together with air-delivered munitions would be much more cost-effective, but let's stay in the undersea regime for now.) What would an ASW operational force look like given these developments? We need to accomplish three operational tasks: detect and localize the enemy submarines or submarine force, establish targeting quality information on the enemy submarines, and attack the detected submarines successfully. We need to do all these tasks while achieving a favorable cost-exchange ratio.

At the start of a conflict, an opponent may have some submarines at sea and others in port. It would be highly desirable to at least approximately geolocate each enemy submarine and keep it under surveillance continuously, from prior to the start of a conflict until the submarine was destroyed. I won't go into the potential to detect or track from space or the air here, but that should certainly be considered and will be discussed as a multi-domain consideration. For now, we'll hypothesize an ability to have multiple hunter-killer UUVs on station to track any enemy submarine,

---

6   Here I'm taking the opposite approach to the sea surface mission focused approach used earlier. Submarines are generally attrition weapons whereas surface ships have historically been used to attempt fleet level decisive battles.

either from the time it leaves port (both prior to and during a conflict), or as it transits a known choke point, and to maintain that track indefinitely. The basic operational organization I envision is a small group of nominally three medium-sized UUVs that can operate together to defeat and destroy an enemy-manned submarine. The unmanned and autonomous platforms needed to do this would be much smaller (and therefore much cheaper) than manned submarines, but they would need the capability to move stealthily at the operational speed of a target submarine. Sensors on this UUV would be primarily passive to avoid detection but could and probably would include active sensors, on at least one of the UUVs, to support engagements. In situations where the enemy submarine's location or even presence was uncertain, one basic scheme would be to have one of the UUVs go active—transmitting sonar signals that would provide bistatic and monostatic information about the threat's location and movement.[7] This acoustically active transmitting UUV would be effectively sacrificial as it would expect to draw fire from the target submarine. All three UUVs would simultaneously engage the enemy submarine, achieving a favorable exchange ratio of no more than one UUV lost to destroy the manned enemy attack boat—a much more valuable and expensive asset than the UUV.

It could similarly be presumed that a friendly UUV stationed near an enemy coast would be destroyed quickly once it had attacked a transiting enemy submarine (or surface ship). If hostilities had already started, then engagements could be conducted as soon as an opponent's submarine attempted to leave or enter port. If not, then the unmanned platforms could remain passive, probably concealed on the seabed (effectively as dormant smart mines), until activated at the start of hostilities. A necessary feature would also be the ability to reliably and clandestinely receive offboard activation instructions at the start of hostilities, possibly to report engagement results, and receive relocation and new mission instructions. Continuous human control is not assumed or required.

---

7   These terms refer to the path the sound waves travel. Bistatic originates with the sender UUV, bounces off the target submarine, and is received by a second UUV. Monostatic originates the same way but bounces back to the original sender.

A concept for attacks on surface ships is even simpler, at least from the UUV perspective. In this case, the UUVs are stealthy weapon carriers that need to be positioned within reach of the planned engagement zone or of known surface targets. The detection, geolocation, and engagement quality track would generally be provided by other assets—space systems or airborne systems or sea surface sensors—and relayed to the weapons on the UUVs. The UUVs with their weapons would then close the loop of the engagement with torpedoes and either cruise or ballistic anti-ship missiles. In operations at scale, say, against a carrier strike group or convoy, a technique to improve cost-effectiveness would be to have UUVs empty their magazines before they could be engaged.

An intermediate concept short of a single or group of UUVs operating autonomously would be to couple multiple UUVs with manned submarines. This concept is analogous to the entry-level use of CCAs controlled by a crewed fighter that I initiated for the US Air Force. It would allow a degree of positive local human continuous control of the formation of UUVs, but it would also put a crewed submarine at risk.

BUILDING BLOCKS

Experimental UUVs roughly in the right size ballpark, such as Orca, have already been built. The Navy refers to them as extra-large unmanned undersea vehicles (XLUUVs). These UUVs weigh up to about one hundred tons and are up to about hundred feet in length. Mine delivery, ocean surveillance, reconnaissance, and intelligence missions are near-term applications.

For lethal applications, there have been recommendations that UUVs be used for unmanned Tomahawk cruise missile launchers and mine-delivery vehicles. Current lethal UUV missions operationalized by Ukraine include attacks on fixed installations and ships in port.[8] The extension

---

8   Somewhat smaller UUVs such as Ukraine's large UUV, Toloka 1000, revealed in September 2025, is reported to have a fixed-target attack mission (for example, a bridge support). In December 2025, Ukraine used a UUV to attack

from limited capability and experimental systems to a collaborative lethal autonomous ASW capability is a matter of integrating the necessary sensors and weapons and command and control capability and gaining confidence in their reliability and effectiveness. This is not a huge leap. Exploring the realism and operational potential for this concept will require multiple prototypes that can operate as a team in the manner described above. For operational assets, there are several design trades that will have to be conducted to balance cost and the range of features needed to implement the operational concept. These problems are not trivial, but the most significant may be achieving the combination of confidence level needed to support engagements and the envisioned range, data rates, and reliability needed for communication among the small unit of collaborating submerged UUVs. Modular designs that enabled multiple mission loadouts would be attractive and improve the utility and cost-effectiveness of the UUVs. This is the intent of the US Navy's Orca XLUUV program already. As in other unmanned systems and domains, losses associated with attrition are expected, so cost will be a major driver. These systems would have to be acquired in significant numbers, but their cost should be a fraction of the alternative of a manned nuclear attack submarine or that of an enemy submarine or surface combatant.

In this concept, the XLUUV serves as the transporter and launcher. The projectile for the ASW would be a torpedo, which would be an evolutionary version of current systems. If justified by trade studies, the UUVs in the concept could be designed to be launchers for projectiles for other missions, ASuW and strike especially, providing some redundancy with other systems for these missions.

---

a Russian submarine in port. "Ukraine Strikes Russian Submarine with 'Sub Sea Baby' Drone," *Naval News,* December 16, 2025,https://www.navalnews.com/naval-news/2025/12/ukraine-strikes-russian-submarine-with-sub-sea-baby-drone/.

COMPLEXITIES

Fundamental Operational Needs

Command, Control, Communications, and Battle Management: As noted earlier, continuous communications with manned installations ashore or on surface ships would be problematic during submerged operations. There are ways to use deployable buoys and other devices to achieve this in some circumstances, but not, as far as I know, during tactical operations. Receiving commands and revised rules of engagement this way is relatively easy, followed in order of difficulty by low-data-rate acknowledgements and status reports from the UUVs. Detailed status reporting and data-intensive intelligence reports are most stressing but can be provided for, possibly after moving to a less threatened environment. This seems to be the Navy's current intent for the Orca program. In any event, both individual and small groups of UUVs would have to conduct submerged operations without continuous human control. I don't see this as a showstopper for the concept, but it is a critical operational need. For tactical operations involving collaboration among multiple UUVs, local undersea communications must be adequate to support tactical behaviors. The same would be true of a co-deployment with a crewed submarine. If one UUV is using active sonar for threat detection, that UUV can use active acoustic signals to inform the others in the collaborative team of its location and provide information on the threat environment. Laser systems would be more secure but are sensitive to the undersea environments and are range-limited. The active UUV platform would take on the role of local central node and all other platforms orient around that node autonomously.

Cross-Domain Considerations: The focus of the concept is ASW, which is a heavily cross-domain mission. Airborne assets play a major role in ASW. Surface ships with associated aviation can play a role as well. I would expect these roles in ASW to continue but transition to uncrewed platforms. I would also expect space-based sensors to play an increasing role in ASW. The envisioned XLUUV assets also have roles in other domains beside undersea. Obviously, ASuW missions would be viable and cost-effective. Submarines have a traditional mission of interdicting enemy shipping—military and commercial—and the XLUUVs could contribute to that mission. Large UUVs move slowly compared to surface ships.[9] As a result, ambushes at choke points or harbor approaches and engagements are more achievable than unconstrained open ocean ASuW. Torpedoes have multi-domain applications, both surface and subsurface, but they are relatively short-range. Anti-ship missiles launched from XLUUVs would provide a much more flexible surface ship engagement capability and improve the survivability of the launch platform but reduce the inventory for the ASW mission. The UUVs in the concept will depend on offboard long-haul communications relay nodes for command and control. Space is the most efficient way to meet this need, but airborne relay options are also possible.

Logistical Support: The UUVs in the concept would have ranges of several thousand miles and long endurance. Orca's reported range is sixty-five hundred nautical miles, and it has an endurance measured in months. As a result, logistics support could be provided from ashore in secure locations, although the potential for replenishment at sea is certainly worth considering. There would be some operational value to provisions for munitions and other expendables replacement forward, given the slow transit speeds expected, but the survivability of the logistics vessels would have to be high enough to justify doing this.

Tactical Mission Variations: Depending on how successful modular designs are, and on the planned loadout of munitions, XLUUVs could perform a variety of functions, most of which do not require tactical real-time collaboration. These include ISR, mine delivery, EW, and possibly cyber. Mission variations that involved alternative payloads would have to be planned well in advance, however. Single-platform or noncollaborative mission assignments are certainly possible, and that's the first capability likely to be fielded. At the other extreme are many-on-many mission

---

9   Orca's reported transit rate is about three knots.

situations, such as ambushing an invasion fleet, that would require large deployments and reward higher-scale collaboration for more efficient target allocation.

Target Identification/Collateral Damage Avoidance: Location and threat classification as a submarine or capital warship would be adequate during hostilities and in an area where there is known to be no friendly or neutral shipping. In those cases, rules of engagement could be relatively permissive. Even in a situation with loose rules of engagement, it would be useful to have target identification, through acoustic signature alone, say, adequate to support autonomous decisions favoring high-value target engagement. In other situations, higher or much higher confidence levels would be needed and would have to be verified through extensive testing and exercises. To a significant degree this can be achieved through careful operational planning and collaboration.

Design Requirements and Considerations

Active Defense (Counter-Torpedo Defense): The XLUUVs in the concept are not small (over fifty feet in length) and valuable enough to consider limited self-defense provisions, including active defense measures and expendable decoys in the design. It's a cost-benefit trade-off worth considering, but the results are not obvious, either from a feasibility standpoint or for cost-effectiveness. If these platforms are intended to be attritable, or if a subset of them are sacrificial, it may be hard for defensive systems to earn their way into the concept.

Anti-Tamper: There is a high probability the UUVs in the concept would fall into enemy hands at some point. Either the wreckage of defeated systems would be recovered, or UUVs could be trapped in some form of disabling netting, including commercial fishing nets (although the Navy's current XLUUV design requirements include a provision to avoid this threat). Anti-tamper provisions would govern all the sensitive systems aboard, especially weapons, secure communications, and autonomous behavior control.

Communication: At the surface or with the assistance of a deployable buoy, state-of-the-art secure radio frequency transmission to space-based or airborne assets would be required. In some environments, commercial SATCOM could be used. Tactical acoustic or optical communication for undersea operations and tactical collaboration were discussed earlier.

Confidence in Automated Behaviors: The undersea environment is a relatively benign environment for transit—there isn't much to bump into as long as one navigates to avoid the seafloor and stays at a reasonable depth. Surface or near-surface operations are similar to those for USVs. Individual XLUUV tactical behavior is relatively straightforward. The greatest challenge will be automated collaborative tactical behaviors, both the coordination of attack tactics and the avoidance of fratricide.

Cyber and EW: The XLUUVs in the concept could conduct EW missions—collection in denied areas especially. They could also serve as a node for wireless cyber insertion into commercial or even military networks. Specialized payloads or modules could be acquired for these purposes.

Deception: Concealing the XLUUVs until they conduct engagements is essential. They should be designed to remain passive until energized by the presence of threat signatures or as ordered. There would be operational value in providing false XLUUVs as both decoys where real XLUUVs are concealed and to very cheaply influence enemy behavior where they are not deployed.

Deconfliction for Engagements: This is a design requirement, but I don't think it is a major problem for the ASW mission. The targets tend to operate individually and are high value relative to the weapons the XLUUVs would employ against them. We can probably afford to accept a modest risk of waste of some weapons in this case, and it may well be preferable to engage a given submarine or even surface ship with multiple weapons. If these platforms were used for ASuW, against an invasion force, for example, deconfliction would be a greater concern in the design.

Energy: UUVs will have to be designed for adequate range to perform their missions, but that can be achieved with current air-independent diesel technology. More efficient sources of energy are likely to become available in the future. It is also conceivable that refueling during an extended mission could take place.

Fixed Detection and Tracking Seabed Acoustic Arrays: They can certainly have a role in the concept as early warning and cueing sensors as they do now. Covert deployment in locations of operational interest is important to deny defeat of these arrays by adversaries. Similarly, the US needs to have the means to locate and disable adversary arrays, preferably covertly. UUVs can be employed, in peace and wartime, to serve these ends.

Fratricide Avoidance: For ASW, local communications with other co-deployed UUVs and preplanned formations and tactics should be adequate to address this problem. Advanced coordination with allies should mitigate any risk to friendly UUVs.

Humans in the Concept: The humans in the concept are remote overseers of the XLUUVs. They provide operational guidance, policy, and force management, but generally not tactical control, from a secure enclave remote from the XLUUV force.

Intelligence Integration and Dissemination: The concept envisions small groups of operational platforms with limited communications operating at extended range for long periods (months). In this circumstance, intelligence integration and dissemination will occur under human supervision at a secure headquarters. The deployed XLUUVs would usually only need low-data-rate inputs. On some occasions, however, updated data files (such as threat signatures) and updates to onboard software (for example, signal processing algorithms) would need to be transmitted to the XLUUVs. This could be accomplished through several alternative means and could involve temporary movement of XLUUVs to more RF-secure locations. In any event, it is a design requirement.

Interoperability: I don't see a firm requirement for the XLUUVs to operate with allies. Avoiding fratricide is a firm requirement but can be achieved through coordination with the headquarters element instead of through direct contact with the XLUUVs. If the allies field similar systems, closer coordination will be necessary for combined operations; the defense of Taiwan comes to mind.

Legal Constraints: There is a range of peacetime and wartime constraints that would have to be addressed, but nothing prohibitive as far as I know. Acceptable target identification to avoid collateral damage is the biggest and most obvious requirement. This function would have to be fully automated (within directed rules of engagement), unless continuous communication could be achieved somehow—something I don't foresee at this point. Once activated to conduct lethal operations, the XLUUVs are essentially mobile smart sea mines. They would be limited by anything that constrains that category of system. There are some restrictions on sea mines in Hague Convention VIII and, under customary international law, intended to prevent indiscriminate damage to civilians and nonbelligerents, but nothing prohibitive.[10]

Manned and Unmanned Teaming: I discussed this possibility earlier as an intermediate step similar to the first increment of CCAs for the air domain to be covered in more detail later. The Navy has also experimented for years with submarine-launched UUVs, either from torpedo tubes or dry deck shelters. The large-displacement UUV (LDUUV) called Snakehead, which would be deployed from a manned submarine and act as a scout for the manned submarine or perform other missions in conjunction with and in support of the manned submarine, was canceled in 2023. However, testing with the existing prototype was recently restarted.[11] The limitations on underwater communication between the sub and the UUV would limit the utility of the UUV. Its detection would also be a

---

10   In the land domain, the US has had limited success with developing Ottawa Treaty-compliant smart mines with humans in the loop to authorize engagements. (The US is not a party but, until recently at least, has accepted some limitations.) The cost-effectiveness of these systems is questionable, and the military sponsors of land mines have had higher priorities. In the sea domain, mines, including smart mines, are not currently constrained by treaty and are in various countries' inventories.

11   Zach Abdi, "US Navy Restarts Snakehead LDUUV Program," *Naval News*, March 22, 2024, https://www.navalnews.com/event-news/sea-air-space-2024/2024/03/us-navy-restarts-snakehead-lduuv-program/.

cue to the presence of the submarine. It's possible that this "scout" could penetrate areas where the sub could not, but the XLUUV should be able to do the same. It can also carry much more payload than a crewed submarine-launched system, as well as operate from secure bases.

Mobile or Transportable Detection and Tracking Acoustic Arrays: These types of systems have been in use since the early days of the Cold War. I would expect work in this area to continue. There are concepts for deployable ASW sensor systems, active and passive, that could be used at choke points or to provide deployable and rapidly reconstitutable linear arrays for cueing when a submarine or large UUV transits. The cueing would enable airborne ASW systems or dormant UUVs to acquire, track, and engage threats efficiently.

Non-Acoustic Signatures: Work has been done for decades on a wide array of sensors other than acoustic. These include magnetic anomaly detectors, chemical sensors, bioluminescence, wake detection, and others. Signatures do exist in each of these and other physics regimes, but the problems of low signal-to-noise ratio and false alarm rates have generally limited their use, although some systems have been deployed by the US and others. As sensors become more capable and processing capacity increases (think quantum sensing and AI-enabled processing), there isn't much doubt as to where we are headed. I believe that submarines as we know them are going to lose their stealth characteristic sooner or later. It's just a question of how soon. I didn't assume that happened in this concept, but once it does, the undersea domain starts to have characteristics more like the surface domain, and the value and utility of traditional undersea systems declines sharply.[12]

Operational Planning and Rehearsal: This is a problem for the concept I've described because of the limited communications options to UUVs and the long lead time to deploy assets on station. This isn't very different from the current situation with crewed submarines, however. UUVs and any seabed systems will have to be pre-deployed or dispatched

---

12   As noted earlier, this event will have serious implications for the
      nuclear triad, a subject outside the scope of this book.

weeks ahead of any anticipated engagements. They would be able to receive updates periodically, and those could include operational planning guidance as needed. Rehearsals would be through simulations run in the controlling headquarters as part of the mission planning effort. For UUVs that would work as small teams in hunter-killer combinations, the algorithms for collaborative engagement planning and attack would have to be developed *a priori* and would be tested in live and virtual environments prior to deployment.

Peacetime Vulnerabilities/Physical Security: The UUVs in the concept, and any seabed sensors or smart mines included, would be at some risk of destruction, neutralization by other means, or adversary exploitation if detected prior to use. The design and operational concepts are intended to minimize the risk of detection, but one has to assume it would occur at some point. For example, a propulsion or control system failure can't be ruled out. There isn't much that could practically be done about in-situ destruction, or even some forms of nonlethal neutralization (nets, for example). Anti-tamper provisions would be required for certain. Anti-access and/or self-destruction devices could be included in the design as well.

Recovery/Salvage: It would be desirable to be able to recover an inoperable UUV, but if the reliability is high enough, adding this provision might not be cost-effective. For many reliability failures, a self-recovery option should be available. Towing by another modified purpose-built UUV is problematic but not inconceivable. In peacetime, recovery could be by a surface ship or tender designed for this purpose. If wartime recovery is desired, it might have to be by a specially designed or modified XLUUV.

Reliability: It has to be high enough for cost-effective sustained operations, which means quite high. I don't see this as a major obstacle to the concept, as existing submarines and UUVs have very high reliability, and the technology to support the needed levels generally already exists. This is confirmed by the Navy's willingness to proceed with serial production of the Orca XLUUV. This does imply a very focused and conscious design effort in this area and enough testing to provide adequate confidence.

The autonomous features of the platform would be embedded in software running on commercial hardware, and the reliability for both should be acceptable.

Requirements Creep: Conscious attention should be placed on avoiding cost imposing marginal or low return-on-investment requirements. The natural tendency in all domains is to add more requirements to the design until the concept crashes from its own weight. A lot of the items on this list are good examples of potential requirements creep. The whole point of the concept is improved cost-exchange ratios over current systems.

Responsive Threats: The immediate response would be to find ways using existing ASW systems, especially air and surface ship-based, to negate the XLUUV threat, even if the odds of detection and the exchange ratios were poor. There would be attacks on the chain of control through cyber, EW, and kinetic means. Ground-based headquarters and C3BM facilities would be targeted if possible. There would also be an immediate change in the operational utilization of threat-manned submarines, holding them back from risk especially. The first priority of an adversary would probably be its own coastal waters. Brute-force efforts to find XLUUVs and negate them would be employed, expanding the use of expendable ASW sensors, both active and passive. There would be an acceleration of any ongoing or potential novel sensor approaches to improve detection and targeting. Finally, one could expect emulation—the building of similar capabilities to those in the concept. At least the first several of these responses should be addressed in the original design of the concept and its components.

Single Points of Failure: Some redundancy could be provided, but for many systems on the UUVs, this isn't possible. The greatest single point of failure threat may be adversary attempts to enter the communication loop to the UUVs and disable them, or worse, take control. There has to be a robust communication design, which would likely include multiple link options, low-probability-of-detection or interference modes, and a resilient low-data-rate mode.

Stealth: UUVs will be dependent on stealth, just as manned submarines are. They should have some advantage over manned submarines due to the smaller size, absence of life-support systems, human waste generation and disposal needs, and overall reduced volume and machinery mass.

Unattended Sensors: They can be part of the concept and could be deployed by the XLUUVs. For some situations, such as outside enemy harbors or at choke points, unattended sensors could be dormant and activated in either passive or active acoustic modes on command or as pre-arranged. They would be an expendable but cost-effective alternative to losing XLUUVs in the same scenario. They could also be used in the open ocean as a cost-effective adjunct to the XLUUVs during ASW operations.

Other Needed Military Functions

Air Defense: I don't see this as part of this concept, but it's not totally out of the question. There have been concepts and experiments associated with air defense weapons that could be deployed from undersea platforms. A conceivable defense against a fixed wing airborne ASW threat or an ASW helicopter is an automated air defense system (sensor and weapon) that could be deployed by one or more buoys to the surface if an ASW aircraft were known to be operating overhead. Of course, this confirms the presence of the UUV in the concept and would provoke specific countermeasures to the air defense system.

Intelligence Collection: This is a traditional submarine force mission. Without going into detail, some of this mission could be performed by the UUVs in the concept, but some of it could not. The long endurance and loitering potential of the XLUUVs make them well suited for long periods of local observation and collection, without the risk of loss of life or detention of US sailors.

Mine and Countermine Warfare: The XLUUVs in the concept could certainly be used as mine-delivery platforms. That's a baseline mission for the existing US Navy program. They could also serve as transports for smaller UUVs for local mine detection, mapping, and neutralization systems.

Oceanography: This is another field in which submarines currently contribute through data collection. This work could also be performed by XLUUVs or other UUVs.

Seabed Warfare: The seabed is increasingly used for communication links, monitoring, sensing, and engagement using traditional moored or encapsulated mines and smart mines, and increasingly for commercial activity, such as extractive industries. To the extent these activities generate targets of military value, they can be attacked by the XLUUVs in the concept.

Special Operations: Submarines support special operations by providing a covert transporter for SEAL Delivery Vehicles and special operators with their equipment. Because they have no provisions for human life support, the XLUUVs in the concept cannot provide this support. It would be possible, at least for short-duration missions, to include a module for this purpose, but I'm not convinced the utility is high enough to justify the cost. If this capability is a firm requirement, then some dedicated UUVs may be required. If delivery and retrieval is the limit of the requirement, it can be done with UUVs and probably with more acceptance of risk than for a manned attack submarine.

Strike Missions: The XLUUVs could be designed to carry cruise or ballistic missiles. Of the two, cruise missiles are probably much more cost-effective because they could be launched from torpedo tubes without major modification of the XLUUV. I don't see these systems as highly cost-effective, but some of them could be acquired as part of a larger force structure. The reason they aren't cost-effective is the overhead cost of the XLUUV as a transporter for these weapons is relatively high when compared to land-based missiles or bombers and multi-role fighters. Weighed against this application is the small number of weapons that can be carried, the risk to the launching platform once it starts firing rounds, the possibility of discovery and neutralization before use, and the very long deployment and resupply times compared to bombers, for example. This capability can have high operational value in some situations by destroying a few high-value targets in the opening stages of a conflict, as a raid, a prepositioned deterrent, or an enabling step for some follow-on operations. I wouldn't rule this out, but it will take some effort to justify it, especially at scale and relative to other options.

CONCLUSION

Classification concerns severely limit how much one can discuss subsurface warfare. At a high level, however, there is a window of opportunity before the oceans become transparent, metaphorically at least, in which large unmanned undersea vessels can make major warfighting contributions. That day may well still be some years off and can be extended by reliance on uncrewed systems. At this time, the potential for the oceans to become transparent should not constrain us from moving forward with innovative unmanned undersea systems and concepts of operation.

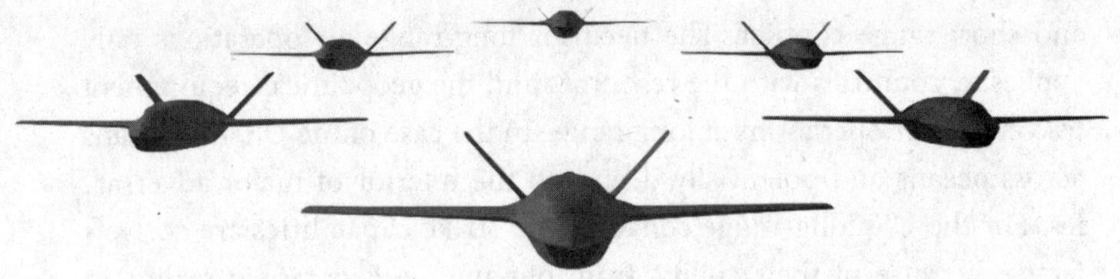

CHAPTER 5

WARFARE DOMAINS– AIR

As the Secretary of the Air Force from 2021 to 2025, I had the opportunity to direct the course of Air Force and Space Force modernization.[1] This chapter and the following chapter on the space domain reflect the decisions I made as the Secretary and the direction in which I led the Air Force and Space Force. Military modernization is a never-ending journey and a race against potential adversaries. I did my best to point both services in the right direction and move them forward as quickly as possible. This chapter and the next provide updated versions of the thinking that motivated my decisions.

In the US, we need to think about the air warfare domain as two separate subdomains. One is the short-range air domain (think in terms of hundreds of miles), and the other is the long-range air domain (thousands of miles). As a global superpower, the US is one of the few nations whose Air Force is designed around these two sets of capabilities. There are compelling reasons for the US to have air forces structured for long-

---

1   As I was writing the original DARPA report that forms the basis for this book a few years ago, I had no idea that I would have the opportunity and the honor to serve as the Secretary of the Air Force for almost four years.

and short-range conflict. The need for long-range air operations only applies to countries with the resources and the geopolitical requirement to conduct air operations at long range. In the case of the US, this means across oceans and potentially deep into the interior of major adversaries. For the US, long-range conventional strike capabilities are cost-effective because of their ability, from our unique geographic reality, to flexibly defend our global interests against any threat.[2] If we were only worrying about defending the continental United States and having some retaliatory capability, we could limit ourselves to local air defenses and long-range offensive systems, but that isn't the case. We have close allies who have no option geographically but to engage in the short-range air domain. Our overseas allies do not have the favorable geography the US enjoys, with friendly neighbors and oceans to protect us. We also have international interests located near potential adversaries that we need to protect. Finally, if we are going to conduct offensive ground operations or sea control operations anywhere in the world, we would need the ability to operate cost-effectively in the tactical shorter-range air domain environment where adversary ground and naval forces operate.

In both long- and short-range air subdomains, we also need to include air and missile defense. I include ground- and sea-based air defenses, except for close-range ground force or sea surface tactical unit self-protection systems, as part of the air domain forces. Earlier, I primarily discussed air defense in land and sea domains in the context of force protection. Here, wide-area regional and national air and missile defenses are a significant part of the struggle for control of the air domain itself. Historically, the US hasn't emphasized homeland or theater-level surface-based air defense the way our potential adversaries have, but that doesn't mean we shouldn't or won't be forced to in the future. Ostensibly, the

---

2 The strikes on Iran in August 2025 and in 2026 illustrate the
  flexibility and value of this almost unique global strike capability
  achieved through long-range bombers like the B-2.

Trump administration's Golden Dome program will include significant ground-based air and missile defenses within US territory.[3]

In modern times, the US has recognized the importance of airpower in both short- and long-range air subdomains and has emphasized investments in both. This has been true for decades, ever since World War II. Air systems come at a high cost, however. As Len Sullivan, the former Director of Program Analysis and Evaluation in the Pentagon used to emphasize, "Defying gravity is expensive."[4] It can also be decisive. In the future, control of the air will still be essential to success in war, even as the space domain takes on increasing importance.[5] After the Cold War ended, the US continued the B-2 and F-22 programs but drastically reduced the inventory sizes of both, significantly upgraded the Navy's F-18, and initiated the program that became F-35.[6] A few years ago, the Air Force initiated the B-21 medium-bomber program.[7] More recently, the Air Force committed to the development of the F-47, an air dominance-oriented crewed replacement for the F-22.

When I became the Secretary of the Air Force in 2021, the Air Force and Navy were working to define the next generation of air dominance systems they would acquire for the short-range air subdomain. At that time, the Air Force was on the path to field a crewed tactical fighter that would replace the F-22. This concept incorporated technology from the Air Dominance and Aerospace Innovation Initiatives that I originated and

---

3   I'm highly skeptical that this program will endure.

4   Len Sullivan's point was that air systems on a cost-per-pound basis are much
     more expensive than land or sea systems—true, but not directly relevant.

5   The inability of Russia to seize control of the air over Ukraine has been a major
     contributor to Ukraine's success at halting the Russian ground campaign.

6   As Deputy Director of Defense Research and Engineering for Tactical
     Warfare Programs, I was a participant in all these post-Cold War decisions.
     I was also "in the room where it happened" when John Deutch, then
     Under Secretary of Defense (Acquisition) made the decision in 1993 to
     continue the F-22 and F-18E/F, stop the Navy's A-12 follow-on, and start
     the Joint Advanced Strike Technology program that became F-35.

7   The B-21 is a medium-range bomber versus an intercontinental-range bomber like
     the B-2 and B-52. This decision was driven by affordability concerns. It can still
     accomplish long-range missions but requires additional tanker support to do so.

sponsored as Under Secretary for Acquisition, Technology, and Logistics.[8] The Navy was moving in the same direction, but more slowly. In addition, the Air Force, and to a lesser extent the Navy, emphasized integrating all their assets into a military internet of things (IoT) in which a cloud-based computer architecture provides highly coordinated data analytics and optimized C2 to the force in real time. The Air Force version of this was called Advanced Battle Management System (ABMS). The Navy version was called Project Overmatch, and the joint version was called Combined Joint All-Domain Command and Control (CJADC2). In 2021, unmanned systems were not a central part of either Navy or Air Force planning. The Air Force had been experimenting with uncrewed prototypes. The Navy had initiated a program to acquire an aircraft carrier-based uncrewed tanker, the MQ-25.[9] Both services seemed open to augmenting manned aircraft with unmanned systems for support functions but not to go beyond that point. Other countries were exploring "loyal wingman" concepts, with unmanned systems operating alongside or in advance of more traditional manned tactical aircraft. These efforts were all well intentioned and generally in the right direction, but as I became Secretary of the Air Force in 2021, I believed we needed to consider some fundamental issues in the air domain that were not being adequately addressed.

First, in the short-range air subdomain, the dependency on a small number of fixed and very targetable forward bases for conventional-takeoff tactical-range aircraft (a few hundred miles) is becoming and may already be untenable. This is true whether the aircraft is manned or unmanned. Fixed land airbases and aircraft carriers operating close to an opponent's coast or border are not survivable enough against current and future threats. This is a development that has been coming for a long

---

8   The former was a roughly two-year study that I tasked DARPA to lead in 2012. The latter is a classified technology program that I initiated in 2015 to develop next-generation technology in an X-plane demonstration program led by DARPA but with Air Force and Navy co-funding.

9   The MQ-25 program was originated by then-Deputy Secretary Bob Work and Chief of Naval Operations Admiral John Richardson in 2015. It is expected to enter production in 2026.

time; it was a major concern during the Cold War, even before the advent of long-range precision conventional ballistic and cruise missiles, to say nothing of hypersonic weapons.

Second, the unit cost of modern tactical aircraft, for both long- and short-range air domain applications, their long production lead times, and their expected attrition rate against peer competitors, are prohibitive of the ability to afford a force that can conduct a long air campaign. In a protracted peer competitor conflict, the US is likely to run out of airframes if it attempts to conduct operations over a period of weeks or months.[10] The F-35 is a good example. I'm a big fan of the F-35; it's a tremendous machine and a major step forward over its predecessors, but only if it can achieve high net cost-exchange ratios with very low loss rates against the threats it will face for the next few decades. That means limiting losses on the ground as well as in the air. If it can be based in a survivable way, the F-35 (with adequate investments in continuous upgrades, EW, C3, air-to-air weapons, and coupled with attritable UAVs) will serve the US well for a long time.[11] Unfortunately, its high cost (including sustainment cost), long acquisition lead times, and basing limitations—except for the Marines' short takeoff/vertical landing (STOVL) F-35B model—will limit its potential to support a protracted conflict.

If these were the only challenges the US already faces, it would be enough, but there's more. China, in particular, has been working to develop capabilities that challenge the US directly in the air. This includes attempts to achieve stealth, improved fire control sensors on tactical aircraft (optical and RF), and especially superior air-to-air missiles of various types and ranges that in some ways exceed the performance of current

---

10   This is what happened to Russia in Ukraine. The prospect of unacceptable attrition
      rates has kept Russian Air Force assets outside engagement zones, and Russia
      has suffered significant losses of aircraft on the ground. Ukraine has been able to
      defend, conceal, and relocate its aircraft well enough to limit losses on the ground.

11   Currently, the F-35 can amplify the value of existing fourth-
      generation aircraft by working with those aircraft in teams. As tactical
      unmanned systems are fielded, the F-35 can be used in conjunction
      with those systems to achieve even higher cost-effectiveness.

US weapons. This last category includes missiles intended to attack stand-off non-stealthy counter-air and strike support sensor and C3 systems like the Airborne Warning and Control System (AWACS) and JSTARS. Even tankers, which traditionally have been able to refuel fighters outside of the range of enemy engagement, are becoming targetable at long ranges.

We don't have time to waste as we sort out the right concepts for the next generation short- and long-range air subdomain solutions, but it is also essential we make the right decisions about the future of the air domain. The future concepts I'll describe below can be thought of as the "force after next."[12] The choices the US makes now and in the near future, including the ones I made as secretary of the Air Force, won't go as far as the concepts I'll describe, but they take us partway there and make the path to that future shorter and easier.[13] Conversely, if we make poor decisions, we will not shorten this path but increase its length and difficulty. One of my decisions as the Secretary was to develop and field at scale (meaning in significant numbers) uncrewed Collaborative Combat Aircraft or CCAs. The first increment of these uncrewed lethal platforms is well on the way to fielding in operational units. I believe the CCAs are an important initial step in the right direction. My provocative conclusion is that the manned "fighter plane" as we've understood it since World War I will ultimately become obsolete. Neither the combination of speed and maneuverability as a design driver nor the emphasis on relatively close-in duels between manned aircraft appears in the concepts I'll describe. Missile maneuverability and speed already dramatically exceed what a human is capable of, and future missile launchers or carriers can have similar characteristics—if they are not manned. Tactically required decision times, especially in engagements involving more than a very small number of opposing aircraft, will eclipse anything a human can ever accomplish. "Alone and unafraid" will not characterize aerial combat in

---

12 The "next" force is the one I initiated as the Secretary. It includes the Uncrewed Collaborative Aircraft currently under contract and moving into production.

13 Force after next, because neither the technology nor the culture is ready to support the degree of unmanned performance that I'll describe.

the future. The systems carrying weapons (launchers) and the weapons (projectiles) themselves will not be alone, and inherent human capacity limitations will be an impediment, not an advantage.

OPERATIONAL CONCEPTS

In the short-range air subdomain, I see no alternative but to avoid as much as possible the long-runway dependency of current or projected crewed-aircraft conventional takeoff and landing (CTOL) fixed basing within range of enemy ground-launched precision missiles. Hardening, deception, rapid runway repair, and defenses (soft and hard kill), if we were willing to buy them at scale, can all improve the situation, but precision missiles will continue to make it highly problematic to operate from a small number of fixed and therefore easily targetable forward bases. Aircraft carriers can at least move back, and one can talk about using their mobility to do tactical "raids," but sustained carrier operations within range of shore-based missiles are equally problematic. An obvious alternative is the one the Marine Corps has adopted: short takeoff and vertical landing concepts like the Harrier and F-35B. This is a current option, not a future of warfare potential, but it does point us in the direction of flexible, proliferated, and concealable basing for the forward air domain assets.[14] If we accept this, what should the missions be for a future tactical short-range air domain force, and how might they be accomplished? Classically, we would talk about missions like offensive and defensive counter air, strike, interdiction, and close air support, as well as a number of supporting functions or tasks like ISR, EW, the suppression of enemy air defenses (SEAD), operational-level or above-air defense, and airspace management. What I would propose for the future is off-loading as much of these tasks as possible to other domains, especially space, and focusing

---

14   The Air Force had concepts for alternative contingency basing, but they had not been seriously implemented at scale when I became the Secretary in 2021. Achieving this was one of the seven so called "Operational Imperatives" that I initiated as the Secretary of the Air Force.

the forward-based assets in the concept on efficiently delivering weapons against air and surface targets of interest, while achieving enough resilience that attrition levels are acceptable. Integration and optimization using AI tools would also be part of a future concept, but focused on enabling sound automated decisions at the operational edge at the tactical level. AI would also be used to support higher-level operational decisions and decisions for the force as a whole, where human interaction is more viable and valuable.

What I envision here for the shorter-range air subdomain is a clean-sheet-of-paper force design, not dissimilar to the one described for the ground domain in some ways. A future tactical air dominance organization with the missions of providing offensive and defensive counter air and strike (including interdiction and support to ground operations and units), would consist primarily of a group of unmanned stealthy tactical aircraft with modest runway requirements serving as weapons launchers. These reusable UAVs would have the core function to operate collaboratively and autonomously in small formations of nominally squadron size (tailored to the operational mission) to deliver weapons against aim points or targets designated by offboard sensors. Targets directly relevant to controlling the air domain would include enemy aircraft, air bases, and air defense units. This force would also deliver munitions against ground or sea surface and subsurface targets of interest in support of operations in other domains. ISR, including targeting for both air and ground targets, would come primarily from a combination of space, standoff air, and some ground-based sensors, but as much as possible from sensors and C3 located in space.[15] Standoff or penetrating unmanned aircraft would provide ISR information as well. Efficiency in the concept comes from relatively inexpensive multi-role unmanned weapons carrier aircraft (launchers) that can operate with acceptable attrition rates, and from the ability

---

15 In the near term, a mix of air- and space-based surveillance is needed. The Air Force E-7 program, which was intended to replace AWACS, would provide that capability and offer increased resiliency and redundancy while space-based capability matures. It would also provide critical C3BM capacity.

of resilient space-based assets to provide wide-area threat coverage and timely targeting quality information. Survivable C2 at various echelons would provide optimized weapon-target pairing and tactical cooperation among sensors, UAVs (launchers), and weapons (projectiles). Some UAVs could also be employed as decoys and threat weapon "magnets" to draw enemy engagements, thus forcing exposure, reducing threat weapon loads, and forcing premature operational and tactical commitment of threat assets. In addition to weapons, some of the tactical UAVs could carry decoys and/or countermeasures, including anti-radiation missiles, sensors, and EW modules.[16]

For air-to-air missions, the short-range subdomain concepts' greatest cost-effectiveness challenge is probably fire control sensors and the allocation of functionality between missile seekers and those sensors. The handoff between fire control sensors and beyond-visual-range air-to-air missiles is a critical parameter for modern air-to-air engagements.[17] This function is currently performed by highly sophisticated tactical radar platforms like the F-35 and advanced missile seekers on systems like the Advanced Medium-Range Air-to-Air Missile (AMRAAM). Optical sensors are taking on an increasing role in fire control as well. While space-based sensors or long-standoff airborne sensors may be able to provide aircraft detection and tracking information, I'm skeptical of their ability to provide the precision fire control information needed to bring missile

---

16   A key to cost-effective concepts is deciding what part of the concept is reusable and what part is expended in use. It's the multi-sortie capacity and resilience against attrition that justify the high cost of current tactical aircraft. Uncrewed concepts can rebalance to lower-cost reusable platforms. The very-low-cost drones that have proliferated in Ukraine are at the opposite extreme in cost and reuse.

17   Air-to-air engagements between fighters in modern times are controlled by the ability of the nose-mounted radars or optical sensors on fighters to keep an adversary in the field of view of the sensor and provide fire control quality tracks and guidance information to an air-to-air missile until the missile seeker can acquire the target and proceed independently. In a one-on-one engagement, each combatant is also trying to avoid providing the adversary with the same opportunity. As the number of aircraft in the engagement increases and the opportunity for collaboration is introduced, things get complicated quickly.

seekers within acquisition range of their targets. Detailed design trades will be needed to determine the optimal allocation of this functionality for any future concept.

Once the need to try to keep every platform and pilot safe as much as possible is out of the equation, a whole new range of tactical options opens up. Operational analysis and simulations of CCAs have confirmed this. Heavy reliance on offboard sensors allows operation from greater standoff range for air-to-air weapons. Of course, adding range will raise the cost of those weapons. For the air-to-air problem in this domain, some sensing and countermeasure capability will probably be needed on the tactical weapons launcher UAVs. Passive sensors, working in bistatic and multi-static modes, would be attractive options to consider from a cost perspective. At this point, I would particularly try to solve the close-in air domain ground attack problem without relying heavily on target acquisition sensors located on the weapons delivery tactical UAV launchers. This might become necessary, but it significantly increases cost. It also means that those platforms, or a subset of them, have to operate close enough to acquire targeting quality threat data on ground targets from operational altitudes, a few tens of thousands of feet and with limited standoff ranges, implying much more challenging survivability features. This is the route to an unmanned progeny of the F-35—a penetrating multi-role aircraft capable of operating independently. If it is possible to rely on offboard space-based ISR plus a small number of penetrating, survivable, dedicated ISR platforms, this would be the preferable course of action. There is a whole new world ripe for exploration here, but the possibilities are quite different from current concepts. As the Air Force moves beyond the first increment of CCA, all these trades should be considered.[18]

Humans would be involved in C2, preferably at the equivalent of wing level, but possibly, and I think more likely, at squadron level, the numbers

---

18 Australia is pursuing a concept that reflects my expectations and intent
    for CCA Increment 2. Wes O'Donnell, "Australia's Ghost Bat: The Drone
    That Spooks China," Wes O'Donnell (blog), October 3, 2025, https://
    www.wesodonnell.com/p/australias-ghost-bat-the-drone-that.

would be small, but multiple shifts would be required to support continuous operations. In the "next" force currently in development, crewed fighters provide tactical control of associated CCAs. Experimentation in simulators over the last few years has demonstrated that pilots in current fighters can control several CCAs simultaneously. For reliable command and control, and because of the much better sensors they provide, I would expect teams of CCAs and crewed fighters to be the preferred configuration for the next several years at least. These C2BM cells, airborne or ground-based, would be high-value targets to an adversary, so tactics to provide resiliency and concealment would be of great importance. For the short-range air subdomain concept, I would expect these C2 cells ultimately to be ground-based, but limitations on communication networks might dictate airborne battle managers. For the long-range concept, they would be land-based and rely on space communications. For redundancy, some command and control might have to be airborne, but with as much standoff range as possible. For resiliency, it would be wise to have both options available in each case. The job of command-and-control cells is to exercise human executive control over operations at the echelon where it becomes both necessary and practical. It can become necessary due to vaguely understood sensitive situations and to the limitations of communication networks, data integration, and force management algorithms. It can become practical where the speed needed for decision-making is relaxed enough that inserting humans is an acceptable time burden. The force concept would also include rapid refueling and rearming—also highly automated—to maximize sortie generation. Human fatigue should generally not be a factor that constrains operations.[19]

For the shorter-range fight, forward air and missile defenses would be mobile and designed to support preferential defense of regional assets,

---

19   For an indication of the degree to which AI and automation will be superior to humans at C3BM, see "Air Force Says AI Tools Outperform Human Planners in 'Battle Management' Experiment," *Breaking Defense*, January 6, 2026, https://breakingdefense.com/2026/01/air-force-says-ai-tools-outperform-human-planners-in-battle-management-experiment/.

both military and other high-value assets.[20] Space-based and standoff airborne sensors would provide targeting and tracking data to support launch on warning or launch based on remote sensing depending on the threat. Operationally, air and missile defense systems would dynamically integrate subsets of sensors and launchers to optimally respond to an attack in real time. Ground-based launchers and sensors would be operationally mobile on short cycle times to enhance survivability. Directed energy is a possibility, but weather, threat hardening, and other factors make this questionable anytime soon against threats other than small UASs. Hardening and deception would be used extensively as would soft-kill (jamming and cyber) countermeasures for defense. I would envision at least a two-layer ground-based air defense arrangement with a higher-altitude regional or area defense coverage plus a terminal defense layer. C3BM and airspace management would be fully integrated, and deconfliction between friendly air- and ground-based air defense would be significantly enhanced over current practice due to the elimination of concerns about human losses through fratricide. In my view, all air domain operations (offensive and defensive counter-air and air defense) should be controlled by a single command using a single highly automated C3BM system.[21]

Both the short-range air dominance and the air defense concepts described should be subject to a host of design trades attempting to get the optimal mix and distribution of features across the elements of the force. These trades must include every design variable, but they will be driven primarily by cost and anticipated exchange ratios and other factors described earlier that determine operational effectiveness. Modeling and automating the many-on-many air engagement operational situation is

---

20   Preferential defense assumes selective engagement, based on
      observation of an attack in progress, of threats to some assets and
      ignoring attacks on others, such as decoys or airfields where there are
      no aircraft, in order to maximize the effectiveness of the defense.
21   The US is almost unique in having regional ground-based air defenses as part of
      the Army instead of the Air Force. This dates back to decisions made in 1947.

particularly challenging, but significant progress has already been made in addressing this problem.

Let's turn to the longer-range air subdomain situation in which a thousand miles or more of standoff is assumed. This is a domain subset, if you will, that most countries will never worry about, but for the US, it is critically important. This is sometimes referred to as the global strike operational environment. An operational force here would have to consist of long-range (multiple thousands of miles) weapon carriers or launchers, and standoff weapons with ranges of nominally several hundred miles or more. Alternatively, penetrating weapon carriers/launchers could be used with shorter range weapons. Defensively, it spans defenses against long-range cruise and ballistic missiles, including hypersonic weapons and penetrating or standoff aircraft as threats. For the offensive mission, the US has relied heavily on penetrating bombers ever since World War II. These bombers carry either gravity bombs or relatively short-range weapons. An exception is ALCM (air-launched cruise missile)-equipped B-52s. Early versions of penetrating bombers used in World War II were not as survivable as their advocates believed, and attrition rates, especially in daytime, were extremely high. Through various measures, the US addressed that problem, notably with low-altitude supersonic flight in the B-1 or stealth in the B-2. The B-21, now entering production, is also following the path of penetrating (and therefore expensive) but highly survivable (reusable over many missions) aircraft, carrying short-range (cheaper) weapons. As the Secretary of the Air Force, I supported the B-21 and advocated for additional production as soon as viable, but I also emphasized the need for standoff weapons and additional survivability in the B-21 family of systems.

There are two sides to the global strike long-range air domain equation: offensive and defensive. Currently, the US only emphasizes the offensive component. This wasn't always the case. Once upon a time, primarily in the 1950s and '60s, the US fielded significant quantities of supersonic interceptors for defense of the continental US. They were designed to interdict intercontinental-range Soviet bombers well off our borders. We also fielded an operational US continental ground-based air defense system with air defense batteries protecting several major US cities (interestingly with nuclear-armed missiles). I'm not suggesting we go back to those concepts from the '50s, but in the world in which space surveillance and long-range conventional weapons (including hypersonic boost-glide vehicles and cruise missiles) are possessed by our potential great-power adversaries, we need to rethink this whole equation from both defensive and offensive perspectives. As China and Russia continue to acquire long-range conventional weapons, including hypersonic weapons, it's useful to think about the military target sets in the US these countries might find attractive. On that list would be aircraft carriers and submarines in US ports, bomber bases in the US, and perhaps critical command and control nodes and some key infrastructure, such as critical industry or the handful of sites that support our information and transportation networks. Ambiguity about the conventional versus nuclear nature of inbound missiles and the risk of nuclear retaliation may be a powerful deterrent to strikes on these targets, either preemptively or after a conflict has been initiated, but I'm not sure we should count on that completely. The Golden Dome program being pursued by the Trump administration is reported to be intended to defend the United States against the full range of threats, including the ones I have mentioned. At this point, details about those threats, the defended assets, and the level of protection have not been made public, but it appears to be focused on defense of population centers, not defense of critical military assets or of the nuclear deterrent.[22]

The system-level design trades for an offensive global strike operational concept are driven by the trade-offs in range, cost, and payload for the weapon carrier or, in our parlance, launcher. Today, the medium-range B-21 bomber is being acquired because Secretary Robert Gates canceled its predecessor, an intercontinental bomber, for excessive cost. Secretary Gates commissioned a "family of systems" review led by Ash Carter, then the Under Secretary of Defense for Acquisition, Technology,

---

22   I'm not persuaded that this is affordable, practical, or the right strategic choice.

and Logistics in the Pentagon. Acceptable cost was as much or more of a driver as any operational consideration or requirement, so an intercontinental follow-on to the B-2 was never seriously considered.[23]

A long-range operational strike capability for the US must have a reach of several thousand miles. It must have adequate force capacity overall to be operationally relevant, which means the ability to deliver enough weapons to matter in a reasonable scenario. The reach is the sum of the reusable weapon launcher operational range (including as extended by aerial refueling) and the weapon range—the launcher presumably doing a two-way trip and the weapon a one-way trip. Intercontinental-range cruise missiles are conceivable and intercontinental ballistic missiles exist, but for conventional warfare, both are almost certainly cost-prohibitive. Use of tankers and aerial refueling is part of the equation and may be more cost-effective than building longer-range weapons carriers/launchers, especially if the tankers are already being purchased for other reasons. For the US, we now have to decide how important it is to attack targets deep inside Russia or China, or to confine ourselves to attacking targets near the coasts or borders. If we are talking about the interior of Russia or China, the weapons/projectiles themselves have to be long-range also (well over one thousand miles) or the delivery platforms need the capability to penetrate air defenses and operate in the interior of those countries. If deep penetration isn't operationally viable or required, then this weapon range requirement can be relaxed, and the design of the launcher is significantly less stressing—think an arsenal plane cruise missile carrier (like a B-52 or a commercial derivative) versus a stealthy penetrating bomber (like the B-2 or B-21). I'm persuadable on this point, but the idea of acquiring a force to execute a prolonged conventional weapon-based air campaign into the deep interior of China or Russia seems prohibitively

---

23 Several years ago, when I was the Under Secretary of Defense for Acquisition, Technology, and Logistics, and about to make the milestone decision to put the B-21 into design for production, I asked Bob Work, who was Deputy Secretary of Defense, if he wanted to reconsider whether an intercontinental-range bomber was a better alternative. He chose not to revisit the requirement.

expensive, at least as a peacetime investment. It also raises serious questions of nuclear stability. In my opinion, the stressing military operational need for conventional global strike is realistically more likely to be driven by the need to operationally defeat some form of military aggression conducted on the border of China or Russia than it is to conduct a conventional strategic bombing campaign deep into those countries.

If the US retains all three legs of the nuclear triad (the most likely outcome), then some penetrating conventional strike capability will exist in any event through dual-use systems. This inclines me toward a conventional global strike concept built around a non-penetrating standoff weapon launcher, carrying weapons with adequate range to be released outside most threat land-based air defenses, and weapons with the ability to penetrate to operationally important targets, nominally projectiles with several hundred miles in range from the launcher. That takes care of the problem of getting weapons to targets, but this is just part of a global strike concept. We must address ISR for target acquisition, the weapons mix, details of the launcher aircraft, and C3BM.

To complete the concept, space-based ISR again plays an important role. The problem is primarily ground or surface targets, a much more tractable problem than airborne threats from a tracking and fire control perspective at least. Penetrating survivable airborne ISR would provide redundancy and resilience. Because of the timelines involved for using long-range aircraft with standoff weapons against mobile surface targets, there must be updates during the operations while both weapon carriers (launchers) and weapons (projectiles) are in flight. There are cost-effectiveness trades to be conducted to optimize both weapons/projectiles and planes/launchers and a wide range of possibilities that require quantitative analysis. Hypersonic weapons may be cost-effective for some targets. Autonomous target acquisition by sensors on weapons or on dedicated loitering expendables may also be of high value. Range extension expendables, such as DARPA's LongShot program, may be cost-effective to extend the range of existing weapons.[24] Stealth, defensive aid suites, and

---

24   "LongShot," DARPA, https://www.darpa.mil/research/programs/longshot.

countermeasures are part of this trade study as well. To protect the long-range launcher platforms, even with significant standoff, some defensive counter-air functionality would be desirable. My crystal ball isn't adequate enough to hazard a guess as to how the details of a full concept study would come out. Such a study should consider the scenarios and target sets of interest and likely adversarial responses; the details matter, and a mix of capabilities would be needed in a force designed to conduct continuous strike operations over a reasonable period against a capable adversary.

BUILDING BLOCKS

In both short- and long-range cases, the design trades for weapons and uncrewed aircraft should be conducted without the usual constraint of imposing past design specifications of one element of the concept on others. The US has generally either designed new platforms to accommodate old weapons or new weapons to be carried on old platforms. Given responsive and capable threats, new designs should not be overly constrained by legacy designs. and growth should be built in so upgrades to higher performance are accessible.

For the short-range concept, the UAVs or CCAs are attritable and not full-up fighters. They would likely be subsonic to reduce cost, and maneuverability would probably not be emphasized.[25] Their mission is to survive well enough to efficiently bring weapons—standoff air-to-air and air-to-ground missiles primarily—into the fight. They would have sensors and engagement planning only to the level needed to augment the space-based and standoff airborne ISR and C2 capabilities resident in crewed fighters and/or in dedicated C4ISR (Command, Control, Communications, Computers, Intelligence, Surveillance and Reconnaissance) aircraft in the

---

25　I would use this as a starting point and then explore the value added of more aggressive requirements. It's quite possible that the increased survivability would justify the investment in speed and maneuverability, but I wouldn't assume that to be the case.

concept. Unlike current fighter designs, some subsystems would only be present on a subset of the platforms, considerably reducing total cost. They would be able to act cooperatively autonomously in small groups (like the land concept components) to maximize exchange ratios. The combination of low cost, simplicity, stealth, range, magazine capacity, modular subsystems capacity, and reuse would be the subject of design trades, but the overall intent is to keep the cost low and put in at least enough survivability to get to missile launch ranges for both air-to-air and strike. Each UAV or CCA would carry on the order of six weapons (in some combination of air-to-air and strike).[26] The missile designs would vary based on the target and the sophistication of the threat, but they would have to be designed to overcome terminal defenses, EW, directed-energy countermeasures, decoys, and threat maneuvers. Weapons with loitering and independent autonomous targeting for high-value ground sea surface targets would be included. Small groups of both UAVs and weapons in the concept would be able to autonomously collaborate on tactics against known and expected threats.

For the long-range global strike concept, the standoff launcher would likely be an intercontinental-range aircraft with the capacity to carry on the order of twenty or more total internal and externally stored standoff weapons. The strike missiles would have adequate range to provide at least several hundred miles standoff, and they would have the ability to work together tactically to maximize overall attack effectiveness. Hypersonic capability might be part of the weapon mix, but I'm not convinced at this point it would be cost-effective. Concepts with sub-munitions, various levels of lethality, and variants with features designed to defeat certain target sets would be needed. For defensive air-to-air engagements, passive and perhaps bistatic receiver functionality for onboard targeting and collaborative engagement using offboard sensors and C2 could be included if cost-effective.

---

26 Past studies have indicated that this number is roughly the optimum for air-to-air applications—less is insufficient and more doesn't add much utility.

In the long-range concept, the launchers are basically buses or trucks delivering weapons to a launch point. They need to be reusable to be cost-effective and therefore must be survivable, but with enough standoff, survivability features can be limited. The planes I envision are not small, as there is efficiency in carrying larger payloads. Given that, similar to capital ships previously discussed, there is no major cost payoff to keeping humans out of the long-range strike aircraft because the delta in weight and cost for having them aboard isn't excessive. (Adding a human to an unmanned aircraft invokes a penalty of about one thousand pounds.) On the other hand, there may be no major benefit to having human crews in all the long-range air domain launcher aircraft. At this point in time, I'd be biased toward an optionally crewed concept for long-range standoff weapon delivery so that local formation command and control would be enhanced, as in other domains. Without human crews on many or most platforms, tactics that intentionally sacrifice some systems for an operational advantage are more acceptable and can be considered. The alternative of having a more survivable penetrating weapon carrier or launcher doesn't seem attractive. This alternative would allow for shorter range and therefore lower cost, mass, and volume munitions, but if the force's role is to defeat China's or Russia's power projection or aggression where the target set is the attacking force, the non-penetrating standoff weapon launcher is probably the more cost-effective approach.

As part of the seven Operational Imperatives efforts that I launched as the Secretary of the Air Force, we originally included the idea of a CCA that would accompany and be controlled by long-range strike platforms—bombers or possibly arsenal ships. We didn't find a compelling argument or operations analysis-based payoff for these platforms and dropped them from consideration. During my tenure, we did conduct experiments with using transport platforms such as the C-130 to deliver existing standoff munitions.[27] Those concepts seemed to be realistic and cost-effective,

---

27 "Special Ops C-130 Tests Pallet-Dropped Cruise Missiles in the Arctic," *The War Zone*, November 9, 2022, https://www.twz.com/special-ops-c-130-tests-pallet-dropped-cruise-missiles-in-the-arctic.

assuming the aircraft could be made available and long-range air defense threats effectively negated.

In both long-range and close-range air subdomain operational concepts, ISR and targeting quality information could come primarily from a network of space-based sensors in survivable and reconstitutable architectures. These sensor systems and associated C3 systems would be used to acquire targeting quality information to support global strike as well as close-range operations—against both airborne and surface targets—and to support defensive measures. A fallback or adjunct to those capabilities and a source of redundancy (always operationally attractive) would be survivable unmanned penetrating ISR aircraft. These systems would support real-time targeting and retargeting for close-in and standoff weapons and launchers. For counter-air operations in particular, fire control sensors would probably be needed on the launcher platforms but not required on every launch platform.

As noted above, the dominant fire control sensors for decades have been nose-mounted radars on tactical fighters. More recently, passive optical sensors, infrared search and track (IRST) systems, have become important as a counter to radar stealth as their resolution and accuracy have improved. In addition to fire control radars and IRST sensors on tactical aircraft, the sensors that provide the seekers (optical and RF) on air-to-air and air-to-ground missiles have enabled long-range beyond-line-of-sight engagements and the potential for an engagement from one platform to be controlled from another. Finally, offboard wide-area sensors on standoff platforms like AWACS and JSTARS,[28] and now potentially space-based, can provide target detection, tracking, and fire control support. The sensor and C3BM architecture in both the long- and short-range concepts is equally or more important than the platforms and

---

28   These large sensor programs on commercial aircraft are very high-
value targets. During the Cold War, we calculated that the Soviets
would expend fifty to one hundred fighters trying to reach and destroy
each of these aircraft. Their importance to the outcome of the fight in
Europe, both on the ground and in the air, was significant enough that
the Soviets would have willingly accepted this exchange ratio.

weapons designs. The various sensors are all interconnected and interrelated, and their characteristics should be traded off with other elements of the concepts. Just to add even more complexity, the sensor trade-offs are also influenced by electronic warfare provisions, signature management, and the potential for countermeasures of other types.

Turning to ground elements of the air domain picture, the air and missile defense systems for the short-range concepts would be largely uncrewed, provide preferential defense, be highly mobile, employ deception and countermeasures, and utilize offboard ISR and fire control in conjunction with organic capabilities. A multi-layered, or at least multiple-shot (shoot-look-shoot engagement sequence) architecture would be highly desirable.[29]

Air and missile defense capability to defend against an opponent's global strike capability would require space-based ISR and fire control sensors and long-range ground-based interceptor missiles. The Golden Dome system being pursued by the Trump administration appears to include space-based interceptors as well.[30] This is a tough mission to make cost-effective because of the geometry involved and the options the offense has to avoid or defeat air and missile defense systems, but some capability could add uncertainty, provide a level of threat attrition, and enhance deterrence. Currently, potential adversaries have a limited long-range bombing capacity and no known fielded long-range conventional intercontinental missile threat (hypersonic, cruise, or ballistic).[31] But that is likely to change given ongoing adversary research and development pro-

---

29   A promising technology that emerged during the time I was the Secretary of the Air Force was the Hypervelocity Ground Weapon Systems. John Keller, "Army Putting Pieces Together for Cannon That Fires Hypervelocity Projectiles for Battlefield Air Defense," *Military + Aerospace Electronics*, July 23, 2024, https://www.militaryaerospace.com/sensors/article/55127689/hypervelocity-cannon-air-defense.

30   I was heavily involved in the Reagan SDI program. Technology has advanced for both offense and defense, but the fundamentals of missile defense have not changed.

31   China conducted a flight test that demonstrated the potential for an intercontinental hypersonic weapon (conventional or nuclear) in 2021. At this point there is no public information indicating that China intends to field this capability.

grams and the existence of tempting high-value conventional warfighting targets in the US.[32]

COMPLEXITIES

Fundamental Operational Needs

Command, Control, Communications, and Battle Management (C3BM): Decades ago, a former Chief of Staff of the Air Force made the comment to me that airpower should be delivered like a violent and sudden thunderstorm, not like a slow falling rain. C3BM is what makes this possible. Both the short- and long-range air domain concepts require secure C3BM that supports human-involved operational level planning and largely autonomous tactical execution. In both short- and long-range air domain cases, tactical decisions are made forward and automated. The C3BM architecture must support human decisions, made with AI support and augmenting automation, for a group of assets to be tasked to conduct an operation focused on a set of targets or potential targets in an operational block of air, sea, or ground space in a given window of time. Once those assets are committed, the group of assets is largely on its own conducting the operation. Humans would monitor the operation (presumably one of many) and could intervene with fairly low latency if necessary. Data flow and processing are kept "forward" at the "edge" as much as possible to limit demands on the C3BM system. Space assets and possibly airborne relay nodes play critical roles in the architecture as do directional and secure communication links.

---

32   A surprise conventional strike against continental US-based power projection systems, such our aircraft carriers in port, our long-range bombers on the ground, and possibly key logistics systems, including underway replenishment ships in port or key military port facilities or transport aircraft and tankers on the ground, could have a decisive impact on the ability of the US to respond to aggression. Arguing against this type of attack is the risk of initiating a nuclear war if the attack is perceived as being nuclear. We have no experience with the escalation risks inherent in a large-scale conventional war between nuclear powers.

Logistical Support: The short-range air subdomain concept requires distributed logistics nodes for rearming and refueling and as many viable or perceived viable basing locations as possible. The logistics infrastructure itself has to be distributed and designed with a combination of hardening, dispersion, and deception to avoid destruction. It must include transportation to support the distributed operational concept. Unmanned systems for ground transport, prepositioned supplies, and resilience are all elements of the concept. The UAV designs should have low maintenance overhead as a result of designs for high reliability, modularity, and for ease of subsystem and component exchange. Humans will be part of the maintenance and logistics support system for the foreseeable future. Many logistics functions and a range of possible complications in support operations in general will require human beings, but their role can be minimized. For the long-range air domain concept, the threat of attack on air bases and the logistics enterprise is less but not eliminated. Note that this applies to intercontinental ranges, not to locations in theater. China has already fielded extensive threats, including hypersonic weapons that threaten assets beyond Guam, for example.[33] Terminal air and missile defenses may be more viable, at least against some threats, but can still be overwhelmed. As a result, logistics operations for global strike missions should be conducted from as many fixed-runway basing options as possible, using commercial airfields and in some cases potentially highways. The options here are finite, and prepositioned fuel, munitions, and spares should be covertly allocated to wartime use points prior to hostilities. The planning system should include a "shell game" for returning aircraft to be supported at bases other than the one they departed from.[34]

---

33 For example, see "China Unveils DF-26D 'Guam Killer': New Missile Threatens US Bases and Carriers in Pacific," *Defence Security Asia*, August 25, 2025, https://defencesecurityasia.com/en/china-df26d-guam-killer-missile-threatens-us-carriers-bases-pacific/.

34 I'm envisioning a future in which our adversaries field their own global strike capability, probably with greater emphasis on long-range land- and sea-based missiles. There is no fundamental impediment to this that I can see, and they are already fielding systems for intermediate ranges.

Multi-Domain Considerations: The air domain concepts provide effects and support into the other domains. When cost, time, distance, reusability of the transport vehicle or launcher, and area coverage flexibility of launchers and projectiles collectively are considered, air delivery overall is a very cost-effective way to get munitions to operational level sets of targets.[35] As a result, the air domain assets will be conducting operations against targets in the other terrestrial domains—land, sea surface, and sea subsurface. This fact makes control of the air domain essential to victory in other domains. Because of expanded line-of-sight area coverage, airborne nodes also provide the first or second most cost-effective option for wide-area communication and surveillance, with the space domain as its competitor for first place. The air domain, therefore, plays a critical role in ensuring C3BM and ISR capabilities for all domains are resilient. Airborne assets can also play key roles in EW and cyber for the same reason. Finally, air transport provides the quickest and most responsive, but not the most efficient or necessarily cost-effective, way to transport assets for use in other domains. These considerations are enduring and will drive the need for aircraft for a variety of military purposes. However, those aircraft will be increasingly autonomous. The air domain concepts have dependencies on space for PNT, ISR, and secure communications for C3BM. While some of these dependencies can be mitigated through redundant air domain functionality, space (if available) is a much more efficient home for these capabilities.

Resilient Basing and Ground Operations: If an air base isn't survivable and can't be kept in operation well enough to sustain continuing flight operations, it isn't of much use, and the aircraft at the base may be destroyed on the ground or effectively neutralized by being stranded. Operating tactical aircraft is a high-overhead enterprise. It involves fueling and arming, maintenance, and battle damage repair. Currently, it involves support functions for the humans involved in these tasks. Each of these activities has a physical footprint, as do command and control func-

---

35   For shorter tactical ranges on land (a few tens of miles, say),
       ground-based artillery is most cost-effective.

tions and flight operations functions. Almost any of the items on these lists can be a single point of failure for continuing operations. Protecting a targetable infrastructure against large raids of precision and smart munitions is problematic, to say the least. Against smart peer competitors, the only hope is a full array of defensive measures. That would include hardening, deception, rapid runway repair, redundancy, and both soft-kill and kinetic air and missile defenses. I don't believe this can be accomplished well enough to sustain operations at our current known fixed forward bases in Asia or Europe. It is absolutely clear to me our potential adversaries are acquiring the capabilities they need at the outset of any hostilities to neutralize these bases with high confidence. For reasons stated earlier, I do not believe the US can abandon tactical-range air domain operations. We have to design for and acquire alternatives to vulnerable, easily targetable fixed forward basing.[36]

Tactical Mission Variations (Offensive and Defensive Counter-Air, Strike, Close Air Support [CAS], Suppression of Enemy Air Defense [SEAD], and Interdiction): Both the short- and long-range concepts can perform the range of tactical air missions, with some limitations. Future offensive and defensive counter-air engagements should be dominated by beyond-line-of-sight engagements. The counter-air engagement equation is about seeing and shooting first and having enough of an advantage in reach, missile time of flight, maneuverability, acceleration, and stealth that an air-to-air missile-launching aircraft doesn't suffer mutual destruction with the engaged target from a similar enemy missile fired before the enemy aircraft's destruction. Part of the idea for the short-range concept above is to make the aircraft's cost low enough that even a one-to-one

---

36   After Sweden and Finland joined NATO, I had an opportunity to visit both countries and see how they had designed their Air Forces to survive given their close proximity to Russian missile threats to traditional airbases. Both provide good examples that the US could emulate. Sweden's Grippen fighter was designed specifically for remote operations. Finland has a similar concept for basing and operations and has determined that the Air Force version of the F-35 is compatible with their remote distributed basing concept. Ukraine is doing the same with their air assets. It's just a matter of making the necessary investments.

exchange ratio is still favorable on a cost basis, but the concept doesn't depend on that. If offboard targeting of air threats can be accomplished from survivable assets in space, or from standoff airborne sensor platforms, then air-to-air missiles can be launched at maximum engagement ranges. The potential for multi-UAV collaboration in the many-on-many or few-on-few air combat situations also provides the opportunity for aggressive tactics that accept some losses in return for overall advantage. For example, if the short-range concept is operationally responding to a large-scale attack, it should have a target rich environment, and an individual UAV can be exposed with the expectations it will be able to launch successful attacks on multiple targets based on offboard fire control before being destroyed. This cycle can then be repeated. In the long-range concept. If enemy defensive counter-air was a viable threat, similar tactics could be employed, possibly with even longer-range standoff air-to-air missiles. Strike operations in general would be enabled by offboard targeting and autonomous seekers on standoff weapons, which would enable autonomous ground or sea surface engagements. Both concepts could provide strikes in support of forces in contact or forces within a few tens of miles of contact (CAS and interdiction as I'm defining them) based on offboard targeting and or ISR adequate to release weapons with autonomous seekers and appropriate automated rules of engagement.

Target Identification, Deconfliction, and Collateral Damage Avoidance: For the types of scenarios of greatest concern, the operational context would provide strong or even conclusive indicators of target identification because of operational patterns. In other situations, more stringent control measures can be enabled and, in some cases, remote human-in-the-loop control if it is necessary. In both short- and long-range concepts, inflight updates and changes in rules of engagement should be provided for both weapons and aircraft. Deconfliction is important to cost-effectiveness. For many-on-many or few-on-few situations, collaboration to avoid conflicts and maximize the effectiveness of the attack is an important part of the concept and has to be designed in (an analogy is preventing all the kids on the field running to the soccer ball).

Design Requirements and Considerations

Air Base Survivability: Even for the long- and short-range concepts above, air base survivability will still be an important consideration. The more remote range for the long-range concept does not protect bases from long-range cruise, ballistic, and hypersonic missiles. The short-range flexible and agile basing for the tactical UAV or CCA aircraft in the concept helps, but it isn't a panacea. The US has long neglected active defense, hardening, and deception/concealment in favor of acquiring and retaining larger or newer force structures. I won't dig into why this is true here, but given the threats we have to prepare for, that needs to change.[37]

Airspace Management and Control: This is a broad area that encompasses operational planning, dynamic replanning, deconfliction (not just among aircraft but also for ground fires of rockets and artillery and any directed-energy radiation), identification friend or foe, and collision avoidance. This is essentially what the ABMS and JADC2 programs are trying to enable now. Unfortunately, the US has had several false starts in this area (for example, Single Integrated Air Picture, Air Operations Center upgrades for C3BM, and early versions of the Three-Dimensional Expeditionary Long-Range Radar [3DELRR] ground-based radar) and is at risk of repeating this history. I don't see a technological showstopper to creating a highly integrated, flexible, and responsive system that incorporates AI technologies and a modern communication architecture, but the difficulty of achieving this vision shouldn't be underestimated, and reasonable steps toward the goal have to be well defined, manageable, and contribute operationally in measurable ways.[38]

Anti-Tamper: Aircraft and the things they carry will come down where enemies will gain access to them. They either must have strong anti-tam-

---

37   As the Secretary of the Air Force, I was able to implement these priorities through my seven Operational Imperatives, budget decisions, and organizational changes. That work is well-begun but has a long way to go.

38   This was another goal of the Operational Imperatives. As the Secretary of the Air Force, I created an organization to run this effort for both air and space and put the best technical manager I could find, Major General Luke Cropsey, in charge.

per defenses against reverse engineering and exploitation or features we don't mind our enemies having access to.

Battle Damage Repair and Resilient Design: There is a long-standing design discipline for military aircraft survivability and rapid repair of battle damage going back to the early days of air warfare. Fused fragmentation warheads on air-to-air and surface-to-air missiles can partially damage their targets over a wide range of outcomes. Without people in the aircraft, however, a lot of the motivation for damage-resilient features intended to increase the probability of successful aircraft recovery disappears. Also, with light-footprint flexible basing, the case for battle damage repair overhead is weakened. Even so, it can be cost-effective to include some level of resilience and provide for some repair of surviving platforms that manage to return to base. This should be a conscious trade-off, not a default to lowest cost.

Confidence in Automated Behaviors: Most safety of flight-related behaviors will mature in parallel in the commercial world and can be demonstrated through testing and simulation. Combat-related automated behaviors will be much harder to verify. These behaviors must be operationally effective, prevent fratricide, avoid collateral damage, and minimize redundant engagements. It will also be important to verify the effectiveness of both baseline tactical behaviors and validate evolving tactical behaviors acquired through machine learning.

Countermeasures: Each platform in the concepts may carry a range of active and passive countermeasures (radar and missile warning sensors, electronic warfare systems, decoys, and expendables) based on cost-effectiveness and expectations for the expected adversaries. Versions of these countermeasures for current threats already exist in many cases, and more are always in development. Technology is enabling more integrated and responsive systems across platforms as well as collective automated learning from both training and combat experience.

Cyber and EW: The conflict in Ukraine has demonstrated the importance of not neglecting either EW or cyber offense and cybersecurity. Cyberthreats must be hardened against to a level that provides confidence the platforms and weapons in the concepts cannot be defeated through cyber means. EW is a continuously evolving area. The most recent innovations include cognitive EW embedded in multiuse radio frequency systems that also act as sensors and communication devices. The EW architecture is an important part of the trade space for long- and short-range concepts. It encompasses electronic support measures, self-protection jammers, and standoff jammers. At this point, I couldn't say what mix of dedicated escort, standoff, or self-protection jamming makes sense in either concept, but I would lean toward escort jamming for the short-range concept, where a subset of the UAV systems in a formation would have this dedicated role, and self-protection jamming for the long-range weapon launcher UAVs in the long-range concept. All platforms and weapons should be designed to reduce vulnerability to EW threats as much as practical (such as low-probability-of-detection and intercept-RF designs, and embedded-radar EW countermeasure techniques).

Deception: The fairly standard ground-based concealment, signature management, and decoy concepts apply here, especially to the forward-based short-range concept. The US has spent decades free from the threat of air attack. Our adversaries have long emphasized these areas much more than the US has—this must change. There is also a good potential for deception operationally in the air where false targets or sacrificial UAV decoys can be employed to precede real platforms and draw fire, reducing threat inventory and causing the threat to expose itself. Anything that creates uncertainty and delays or distorts threat reactions should be considered for inclusion in the concept designs.

Default Behaviors—Loss of Contact/Control by Echelon: A wide range of possible situations has to be anticipated, in principle and specifically, and programmed for. There are the obvious return-to-base situations, but others are more complex. Actions if a primary target is not found, weather implications, and default behaviors for loitering weapons are examples. In a many-on-many highly automated operational context, it is not going to be practical to have humans in a remote location access all the needed information to make judgments about every platform or weapon-level tactical decision that becomes necessary hours after a strike formation is launched. More executive-level operational decisions, such as abandoning an attack entirely because of the resistance level, or shifting a formation to another operational area entirely, should be made by human decision-makers with AI support. Overall, I would anticipate the need for some baseline default behaviors and a range of tailorable preset and adjustable options depending on the tactical situation.

Energy: Current energy sources would be acceptable for the concepts described, but as more efficient sources become available, they can be adopted for use. The infrastructure to support refueling at operational tempos would be necessary for both long- and short-range concepts and would have to be resilient to operational threats.[39]

Humans in the Concept: As in the other future concepts, humans have a critical executive-level planning, oversight, and command role. They are also necessary for some support functions that cannot be reasonably automated in the foreseeable future. Humans will also have a role in interactions with other parts of the joint and combined force and with civilians. These are either operating-base functions or performed from a secure mobile C2 facility. I would anticipate the command and control staffing for a wing-size force to be a few tens of people. Operations would be controlled at either the equivalent of a wing level or possibly at the equivalent of a squadron level. For survivability reasons, I would not collocate C2 elements with actual UAV operational locations. Some support people would have to be at those locations to perform any functions that couldn't be automated.

Identification Friend or Foe: The IFF systems used commercially and that have military modes will probably still be with us in some form in the future. I anticipate other noncooperative means would be used to con-

---

39   The Air Force and the Defense Innovation Unit have pursued electric power concepts for some time. I'm open-minded about this, but so far, the technology does not support operational applications at scale, partly because of the lack of recharging infrastructure for locations other than main operating bases where the infrastructure is vulnerable to attack.

firm identity and classification to provide better situational awareness and threat assessment. There are several signatures (active and passive) that can be merged to provide for both friend and foe identification. The automation of this function is crucial for multi-platform engagements in situations where the environment is relatively target-rich and time is a critical variable. Some tactical situations will be fairly unambiguous, if the right data gets to the right place to allow decisions at the tactical edge. Other situations (see pre- and post-merge item below) will be more problematic.

Interoperability: Interoperability is particularly important in this domain because one of the most important contributions the US can make to its international partners is assistance with control of the air and negation of threatening adversary ground- or sea-based forces from the air. The long-range concept should be capable of assisting any ally in any location in operationally useful timelines. To do that successfully, the US needs to obtain and share reliable situational awareness and target identification from and with partners. That information allows US air forces freedom to operate against those targets and maximizes the effectiveness of allied systems. Interoperable information exchange is also necessary to ensure allies do not engage US aircraft by mistake. Human presence in allied command centers and experience gained through exercises (virtual and live) are probably equally as important as technical interoperability.

Legal Constraints: The greatest concern here may be limiting collateral damage to unintended (noncombatant) ground, sea, or air targets. Weapons in the concepts are going to be beyond-line-of-sight fire-and-forget for the most part. In some cases, weapons will be monitoring a ground or sea surface area for hostile targets of opportunity to engage, either independently or in groups. In dense many-on-many situations, the rules of engagement will have to meet legal constraints while still allowing automated and cost-effective engagements. In other cases, where time and the threat situation permit, there should be provision for humans in the loop to minimize collateral damage.

Mission Planning, Including Dynamic Replanning: Although the lead times for short-range and long-range mission packages are substantially different, the planning and dynamic replanning functionality is similar. A mission package of launchers with associated weapons, sensors, and supporting functions, such as EW, cyber, and deception, is created, with human executive supervision enabled by automated modeling and simulation. This is a complex task with a lot of moving parts and done with a great deal of uncertainty about enemy actions and responses. As an operation unfolds, and new information becomes available, replanning and re-tasking should occur on short timelines. Realistic data loads and decision timelines will dictate some degree of delegation to lower echelons and greater autonomy as decision time is effectively compressed. At the engagement level in high-intensity operations, decisions will be highly automated with human monitoring.

Operational Planning and Rehearsal: Automated tools that simulate mission package interactions will be run continuously to plan operations, assess likely outcomes, and compare courses of action. These tools should learn from operational experience to improve their performance. Threat models should be continuously updated. Currently, air operations involve multiple concurrent planning cycles with different lead times and levels of fidelity. I would expect that paradigm to continue, but with much higher levels of automation and more feedback and "feedforward" between different lead-time planning efforts than currently occur.

Physical Security: Against a future peer competitor, there is probably no sanctuary associated with range, but the likelihood of discovery and line-of-sight ground attack, small UAV attack, or short-range indirect-fire attack is much higher for the short-range forward operating elements of the force. The security of the logistics, command, and communications functions must be provided for as well—at both short- and long-range concepts. I would expect that a combination of humans and autonomous systems would be employed for this purpose. The bulk of the effort is provided by automated means, but human presence is needed to deal with more complex situations and to interact with other humans. The Ukraine conflict has demonstrated that small- and medium-sized UAVs are now a major battlefield threat, but one that can be adapted to and mitigated. This is also a threat that is continuously changing and adapting to countermeasures; that is going to continue indefinitely.

Pre- and Post-Merge Situations: In general, both long- and short-range concepts envision engagements conducted from beyond line of sight. While these pre-merge situations might be preferred (they will be preferred by at least one participant, hopefully the US), post-merge confused situations with intermingled friendly and enemy aircraft will still happen. One side may work hard to force this situation and even accept losses to create it. It's especially hard to avoid this for the short-range concept. Individual and small-group collective automated tactics will have to be developed for these situations. DARPA and the Air Force have already made significant progress with this technology. As the Secretary of the Air Force, I experienced the maturity of within-visual-range automated dogfighting AI "agents" firsthand in an F-16 at Edwards Air Force Base. This technology is advancing quickly.

Requirements Creep: Conscious attention should be placed on avoiding cost imposing marginal or low return-on-investment requirements. The natural tendency in all domains is to add more requirements to the design until the concept crashes from its own weight. A lot of the items on this list are good examples of potential requirements creep, if taken too far. The whole point of the concept is improved cost-exchange ratios over current systems.

Responsive Threats: For the short-range concept, the first reaction would be to try to find and target the UAV or CCA systems and/or their logistic support on the ground by any means possible. For ground targets, deception, mobility, and hardening would be emphasized against both long- and short-range concepts. In the air, tactics would be developed to try to improve exchange ratios. I would expect to see similar concepts fielded but with designs intended to outperform those the US had fielded.[40] For example, extended-range engagement of long-range standoff

---

40   This may already be happening as some publicly available information on China indicates. Joseph Trevithick, "Glimpses of China's New Air Combat Drones Emerge Ahead of Massive Military Parade," *The War Zone*, August 17, 2025, https://www.

weapons launchers and sensors would be considered high payoff.[41] Space assets used for targeting and communications in support of both concepts would also be even more important for an adversary to threaten—by any means.

Single Points of Failure: The planning chains and kill chains in the concept should be designed so there is no single point of failure and there is graceful degradation as assets in the concept incrementally experience attrition or are defeated. I'd be especially concerned about the logistics support network for the forward-deployed and -distributed UAV aircraft and about the communication links that provide situation awareness and targeting data to planning nodes.

Stealth: For the foreseeable future, stealth will still very much be part of the equation. Operationally comparable optical and RF (full spectrum) stealth for sensor wavelengths of interest is highly desirable and I believe technically feasible (except for aircraft and missile rear aspects), at least until quantum sensing becomes available. For the long-range concept in particular, stealth should also be applied to the standoff weapons in the concept.

Training, Experimentation, and Testing: Virtual many-on-many simulated operations and engagements are the key to developing advanced tactics and effective integrated mission planning. Operations in the air, with targeting and situation awareness provided from space, are likely to be decisive in a future conflict. The models and simulations that will plan and control these operations have to be based on realistic and exten-

---

twz.com/air/glimpses-of-chinas-new-air-combat-drones-emerge-ahead-of-massive-military-parade; Stuti Tripathi, "Satellite Imagery Shows China's UCAVs Nearing Operational Status," *Business Standard Blueprint*, October 13, 2025, https://www.business-standard.com/blueprint-defence-magazine/reports/satellite-imagery-shows-china-s-ucavs-nearing-operational-status-125101300615_1.html.

41 China, in particular, is working to continuously extend missile engagement ranges against all targets of interest, including tankers and transport aircraft. "Flash News: China Develops Secret Hypersonic Air-to-Air Missile Posing New Threat to US B-21 Stealth Bomber," *Army Recognition*, January 21, 2025, https://www.armyrecognition.com/news/aerospace-news/2025/flash-news-china-develops-secret-hypersonic-air-to-air-missile-posing-new-threat-to-us-b-21-stealth-bomber.

sive training, experimentation, and testing. Combined live, virtual, and constructive experimentation and testing will be the tools to train the AI algorithms that would do real operational planning. Nothing will be more important to success. If we can't demonstrate substantially improved exchange ratios and higher battlefield efficiency for air campaigns through experimentation and testing, we definitely won't achieve them in practice against responsive threats and in the fog of war.

Other Needed Military Functions

Airlift—Strategic and Tactical: The history for the US has been to develop and purchase military-specific aircraft like the C-130, C-5, and C-17 for tactical and strategic airlift. Tankers, like the KC-46 based on the Boeing 767 and still in production, tend to be commercial derivatives. There will be a need for these functions, but over time, they will probably move to unmanned platforms, at least for most of the inventory. Optionally manned may well make sense for these systems into the foreseeable future because, as in other large platforms, the delta cost to provide for human manning is relatively low and provides flexibility for various situations. Given the vulnerabilities of large commercial airports and the need for remote delivery, I see a continuing need for aircraft capable of operating from more austere locations. Fleet sizes and aircraft capacities would need to reflect expected operational scenarios. During my tenure as the Secretary of the Air Force, we explored a new tanker design focused on survivability and chose not to pursue that design yet, for various reasons. In the near term, it may be more cost-effective to add survivability features to current platforms, but at some point, we will have to move to more survivable new designs and concepts for both refueling and transport.

Close Air Support: An artifact of relying on primarily unmanned systems, especially unmanned small lethal UAVs, in the tactical ground domain is that close air support becomes a less contentious issue. All the emotional aspects of defining close air support and decisions about doing or not doing close air support and how much to invest in it (or not) can be approached more objectively.[42] I spent several years chairing a close air support coordinating group in the Pentagon in the '80s and '90s. It was a struggle to get the Air Force to put resources into the close air support mission, which was seen as a low-payoff use of valuable airplanes and crews that could be attacking much higher-priority targets than the enemy ground forces in contact with individual Army units. Of course, Army infantry platoon leaders and company commanders had a different take on what those priorities should be. Without humans in close contact and as much at risk on the ground, and with small lethal UAVs available, the relative priorities for determining how to allocate aircraft strike sorties and optimize weapon loadouts could be much more objectively determined based on desired operational outcomes and relative cost-benefit analysis. The heavy reliance on armed small UAVs in the ground concept reduces the need for close air support from nonorganic air assets, but it doesn't eliminate it. One can imagine scenarios in which quickly massing airpower delivered from CCAs could augment the available ground force small UAVs at the center of the ground domain concept.

Combat Rescue: At some point as piloted or crewed fighter and bomber aircraft are eliminated, the need for combat rescue dissipates as well, but it probably isn't eliminated entirely. There will still be some need for extraction capacity for a range of missions and circumstances, including special operations or covert operation situations. There is also a dual-use value for humanitarian or noncombatant evacuation purposes. I would expect this capability in the future to be provided by unmanned systems, however. It would allow for higher-performance aircraft and more assumption of risk on the ingress portion of a mission particularly.

---

42   When I was Deputy Director of Defense Engineering and Engineering for
      Tactical Warfare Programs in the early '90s, I chaired the DOD Committee on
      Close Air Support. It was painful to lead the negotiations among the services
      over how much the Air Force would spend on this mission, if anything. For
      years I defended the A-10 Warthog, a plane optimized for CAS, as the Air
      Force tried to retire it As the Secretary of the Air Force, I realized it was finally
      time to retire this forty-year-old aircraft, albeit with some reluctance.

Disaster Relief and Humanitarian Assistance: Aviation assets, both rotary and fixed wing, play a major role in disaster relief. That mission will still exist, and the DOD will be tasked to support it, but it will be a collateral capability and not a major driver for the systems acquired by the DOD. The supply function can be highly automated and conducted with the airlift assets discussed above. Other specialty capabilities, such as medical treatment, medical evacuation, services restoration, and anything involving significant interactions with the local population, will still involve humans for the foreseeable future.

Interdiction: This is a core mission of the envisioned air domain forces, for both the short- and long-range air domains. These future Air Force concepts are designed to defeat an adversary's power projection efforts. The principal targets are initially offensive enemy air forces (in the air and at their bases) to secure freedom of action in the air and defeat air attacks on US and friendly forces, and then interdiction against advancing enemy forces (on sea or land), logistics systems and C3BM capabilities that support them, and against ground- or sea-based gun systems and missile launchers that are supporting that advance. Because of the importance of the space domain to operations in all other domains, space-related targets (ground stations, jammers, launch systems, space surveillance, and ground-based anti-satellite weapons [ASAT] systems, for example) would also be high-priority targets.

Medical Evacuation: This capability will be required, and it will still involve human caregivers for the foreseeable future. Given the reduced reliance on humans in the various domain concepts, however, the amount of this capability needed may be significantly reduced.

Mine Warfare: This is debatable, but I don't see the US moving to significant reliance on or utilization of air-delivered mines to support land warfare. In addition to the concerns about collateral damage and disposal, there is a cost-effectiveness concern about integrating land mines with maneuver and fires. Mine use in sea surface and subsurface maritime domains is more attractive, given the threat targets and areas of operation in which sea mines could be employed against potential adversaries. There isn't much in the way of new technology needed to field bottom-moored or emplaced air-deployable smart maritime mines for use against threat surface ships and submarines. Operational concerns about feasibility of deployment, timing relative to threat actions, endurance, command and control, and net effectiveness should be analyzed before going down this path too aggressively. On the countermine side of the equation, there is a potential role for airborne countermine sensors and delivery of countermine sensors and neutralization payloads for sea mines. Both applications could be accomplished by unmanned air vehicles.

Noncombatant Evacuation Operations (NEO): Airlift plays a heavy role in NEO when time is limited and extraction of civilians at risk is the high-priority mission. Commercial air would be used as much as possible, but if conditions dictated, military airlift could be used as well. This can be a complex mission, with a strong need for human presence to deal with security, government-to-government coordination, and just the complexity and unpredictability associated with a situation where a NEO effort has to be conducted.

Special Operations: Specialized air assets, rotary wing and fixed wing, play a major role in a range of special operations missions. Operations like raids, extraction, insertion, hostage rescue, and so on involve a great deal of both detailed planning and rapid adaptation to changing conditions. The complexity and unpredictable features of many of these missions mean there will be a strong human element to conducting these missions for the foreseeable future. UAVs are already used to support these missions, and their use will expand, but they will continue to be tightly human-controlled operations and often involve human operators on the ground for the foreseeable future. That said, technology will provide the humans involved with advanced unmanned platforms and automated tools to assist with planning, rehearsal, and execution of these missions.

Strategic Air Campaigns: The forces envisioned could conduct air campaigns against an adversary's economic infrastructure or other strategic target set using conventional weapons, but neither the long- nor short-range concept was designed with that capability in mind. This can be a practical and effective way to coerce a lesser power, as was done against Serbia in 1999 and attempted against Iran in 2026. However, I don't see a protracted strategic air campaign as a practical reality when dealing with countries like China and Russia, which both have large nuclear deterrents and high capacity to absorb damage while extracting losses through attrition. The risks and the costs of sizing a force for this mission seem prohibitive to me.[43]

Suppression of Enemy Air Defense: This capability will be required and specialized weapons (kinetic and non-kinetic) and tactics will be needed to execute this mission.

CONCLUSION

I believe that the shift to reliance on unmanned lethal platforms in the air domain has begun with the CCA program that I initiated and nurtured through its first few years.[44] Air Force senior leadership support was essential to this progress. Generals Charles "CQ" Brown and David Allvin provided strong leadership and support. At the home of the fighter pilot, Air Combat Command, General Mark Kelly and General Kenneth Wilsbach provided critical support. At one point, I asked some young fighter pilots working on CCA tactics why they seemed so supportive. Their answer was, "Sir, we know these things are going to keep us alive."

---

43   The Air Force Association's Mitchell Institute has recommended that the Air Force acquire 300 B-21s and 200 F-47 fighters for this purpose against China or Russia. "Strategic Attack: Maintaining the Air Force's Capacity to Deny Enemy Sanctuaries," Mitchell Institute for Aerospace Studies,https://www.mitchellaerospacepower.org/ strategic-attack-maintaining-the-air-forces-capacity-to-deny-enemy-sanctuaries/. I don't agree with this recommendation, at least not for this mission need.

44   It was encouraging to learn that in December 2025, the Air Force had added a third competitor, Northrop Grumman, to the CCA Increment 1 program and had selected nine candidates for Increment 2. Northrop had used internal research and development funds to get back into the competition; Thomas Novelly, "USAF Adds Third Contender for Initial Robot-Wingman Buy; Picks 9 for Next Phase," *Defense One*, December 23, 2025, https://www.defenseone.com/business/2025/12/ usaf-adds-third-contender-initial-robot-wingman-buy-picks-9-next-phase/410375/.

They were right. The ideas behind that concept as expressed in this chapter are not mysteries or speculative at this point. The technologies described are either here already or fairly predictable. We are in the midst of an intense race for technological superiority in this domain, which has been at the heart of American power projection since World War II.[45] I would argue, and I will in the next chapter, that the space domain may have displaced the air domain in importance for military success, but we should not underestimate the importance of control of the air and the aggressiveness with which it is being contested today.

---

45   As far as I know, the last time the US fought without air superiority was the Battle of Kasserine Pass in North Africa in 1943. It did not go well.

CHAPTER 6

WARFARE DOMAINS– SPACE

In the air domain, I described two needed concepts, short range and long range. For the space domain, I will discuss two somewhat separable conflicts instead of one—a future offensive conflict, where one tries to eliminate an adversary's space capabilities, and a future defensive conflict, where one tries to defend one's own capabilities against attack. This stems from the nature of space as having military value largely as a location from which to provide support to operations in other domains and from the fact that space systems are generally either one or the other—support systems or weapons—but not both. In a space conflict, there are no fixed territories or zones of control to defend and attack; everything is intermingled. Conceptually then, there are two overlapping warfighting missions in space: providing and protecting the support functions we depend upon space for, and denying similar support to adversaries. Some tools apply to both conflicts, but many do not. Also, because the carrier, launcher, and projectile construct would primarily apply only to the offensive mission and would not simplify the discussion appreciably, I'll forego that awkward intellectual exercise and drop that construct in this chapter.

Space as a warfighting domain today has two other unique features; there has never been a real conflict in space and we are behind China in fielded space order of battle. In other domains, such as sea surface, there may not have been a major peer-on-peer conflict with modern military technology for several decades, leaving us without any relevant current experience to guide us. In space, we have no warfighting experience to guide us at all. We may get the opportunity to have that experience. For decades, China has been fielding a space order of battle designed to support its operational forces and, in particular, to support its anti-access area denial strategy of defeating US power projection in the Pacific. China can easily be considered to be ahead of the US in fielded space order of battle if one looks just at numbers of systems. I've said this publicly many times. We may be ahead in technology, and there is more to the balance of space power than visible fielded systems, but we are behind in numbers of fielded ISR satellites and in the numbers of fielded anti-satellite systems.

In 2014, I showed the Deputy Secretary of Defense's senior budget deliberating body, the Deputy's Management Action Group (DMAG), a PowerPoint slide I had constructed that characterized the choices in space facing the DOD. The slide depicted alternatives necessitated by the fact that China and Russia were both developing a suite of space control systems designed to kill the satellites the US depended on for intelligence, communications, navigation, timing, missile attack warning, and missile defense. Our most important assets in space numbered in the few tens of satellites and were in predictable and visible orbits. For decades, we had operated those assets with almost complete impunity from attack. That time was fast ending if it had not already. My point was the DOD had three choices. We could keep building these kinds of often multibillion-dollar satellites and try to defend them; we could rethink our space architectures and create more resilient or defendable architectures; or we could reduce our reliance on space-based assets. Business as usual was not a viable choice. I pointed out that peacetime intelligence collection requirements—the traditional mission of the National Reconnaissance Office (NRO)—argued for very sophisticated high-resolution sensors, and therefore large, complex, and expensive satellites that are very hard to defend in a conflict. There was certainly an overlap in requirements, but our warfighting needs and our peacetime intelligence needs were diverging. We could either reconcile those requirements or build separate systems for the two missions. I've become increasingly convinced we will be forced to do the latter; exquisite peacetime intelligence collection systems are needed, but they will go on being small in number, big in size, high in cost, and vulnerable to attack. They can be hardened and defended to a degree, but at the end of the day, they will not be adequately survivable against a determined, capable, and well-resourced adversary.

In about the same period, roughly 2014 to 2016 we were also trying to define the content of the Third Offset Strategy concept that Deputy Secretary Bob Work had initiated. As part of that effort, I chartered a Long-Range Research and Development Planning study chaired by Stephen Welby, Assistant Secretary of Defense for Research and Engineering. One of the critical questions I asked that group to address was whether we should continue to rely on space for the important military support functions it provided, or shift to other domains and assets, such as standoff or penetrating long-endurance unmanned airborne systems. The group's answer (which I accepted at the time, but which I was not 100 percent sure was correct) was to increase our reliance on space systems rather than decrease it. To do this, we had to design space architectures that could survive attack and/or be quickly reconstituted at reasonable cost. That was the course of action recommended by the study. When I became Secretary of the Air Force, I found that the DOD was moving in that direction, but still had not fully resolved the issues of just what the US space order of battle should be, how much resilience was needed, or how it would be achieved.[1] In 2021, we had a Space Force and a Space Development Agency, but we had not answered some fundamental questions: What will war in space look like in the future, and what order of battle should the US invest in to

---

1   The first of the seven Operational Imperatives I used to focus analysis and develop budget options was "Space Order of Battle," by far the most comprehensive item on the list.

prevail?[2] After having the opportunity to lead the Space Force for several years, I feel that we have made a great deal of progress. I would particularly credit General John "Jay" Raymond, the first Chief of Space Operations (CSO), and the Space Warfighting Analysis Center led by Andrew Cox for the professional judgment and analysis that informed decisions about creating a new space order of battle for the United States. That work isn't finished, but it is being continued by CSO Chance Saltzman and Secretary of the Air Force Troy Meink. It has happened slower than I wanted, and the rate of progress and funding are not what I would like to see, but we are going in the right direction, and we are beginning to field the essential operational capabilities I will describe here.

In addition to the support functions the United States has come to depend upon, we will also need an offensive space capability. While it is highly desirable to maintain US space-based capabilities in a conflict, it is essential to deny them to threat nations. A peer adversary cannot be permitted to retain access to persistent targeting quality data to support operations. This would put the entire joint force and our allies at unacceptable risk. I have a hard time imagining a future great-power conflict in which achieving control of space would not be decisive. For a country like the United States, whose military is built around the ability to project power and to defend far forward, being able to decisively deny an adversary observation from space is as essential as control of the air has been to ground and sea operations for the past several decades. For this reason, offensive counter-space systems should be a priority investment for the US. It may be possible to limit at least some offensive capabilities of a potential adversary through international agreements, by banning prepositioned co-orbital anti-satellites or nuclear anti-satellites for example. I'm skeptical this can be accomplished, but the potentially hair-trigger nature

---

2   As an aside, when one creates a Space Force, one can expect that institution to see more spending on space as the solution to any threat to space systems, rightly or wrongly. We've created a new bureaucratic and political power center that will advocate for full funding and increases to its own budget. The need for such a power center, given how the Air Force historically prioritized space investments, was probably the most compelling argument for creating the Space Force.

of war in space and the perception of a first-mover advantage suggests that effort should be undertaken. I'm reminded of the naval arms races that in some part led to World War I and World War II. It would be wise to avoid going down the arms race path, but we have already started, and the pace is accelerating.

Policy people like to talk about space being crowded, congested, and contested. Military operators like to talk about it as a warfighting domain and as the "high ground" in future conflicts. I'm inclined to think of space more as a sort of no-man's-land. Both sides have it under continuous observation, both sides can cover it by fire, and there is limited opportunity for concealment. During a conflict, space becomes a hard place to survive, especially over time. Passive systems have a better chance to survive than active systems, but to be useful, almost all space systems must be able to receive and send information. In peacetime, space is open to all and there is high opportunity for unimpeded deployments and deception. This fact grants a potential adversary an opportunity to conduct reconnaissance, gather intelligence covertly or overtly, station sensors and weapons in orbit, and prepare the battlefield in detail prior to the start of hostilities. Presumably, deception would be used to achieve some objectives surreptitiously. This situation is tactically and strategically unstable; it can make the risks of not moving first to attack the other side appear very high during a period of tension. Think of space as a chessboard where both sides get to preposition pieces prior to the start of the game, without full knowledge of the other side. Then either side can start the game at its own discretion by moving all its pieces at once. The rewards for going first, and the penalties or risks for going second, are very high.

All of this is a new dynamic and one that has not been explored or thought about very well at scale. Most war games and scenarios I've seen concern themselves with limited-scope and short-term attack scenarios—more like sniping or skirmishing than a full-out assault. The reality, I'm afraid, is that an adversary would move quickly and decisively, with surprise, if possible, to completely destroy all relevant adversary space-based capabilities. (Note that destroying capability is not the same as destroying assets. There is a high payoff to identifying and destroying key nodes, such as unique C2 assets, that can bring down an entire suite of capabilities.) Ground stations and launch stations would likely be attacked as well, especially the more accessible overseas assets if there was a desire to avoid "homeland" attacks. The attack would likely blend all available types of attack, including kinetic, directed energy, cyber, and EW.

Future space conflicts will be characterized by a high degree of automation and autonomy. Space systems are already going to be unmanned and largely autonomous, especially once they are committed to a course of action. High-fidelity war game participation has taught me time is a critical parameter during a conflict in space. Orbital transfer times, ascent times, fuel burn times, and the tyranny of orbital mechanics conspire to limit operational choices very quickly as decision time compresses. In this construct, effective automated decision-making for battle management can be a crucial determiner of outcomes. Humans may be able to monitor and override automated battle management tools, but those tools and their ability to successfully simulate many-on-many engagements and optimize actions will be essential to success—for both defense and offense.[3]

There is another destabilizing offense-defense asymmetry here—the attack planner may have infinite time to plan and rehearse an attack. If a conflict begins by surprise in space, the attacker can choose the most ideal time for the attack from several perspectives. The defense does not have these advantages; it must respond in real or near real time and do so at a moment the attacker has chosen as the least favorable for the defense. Because a successful surprise attack in space can also be decisive for operations in other domains, the temptation to launch such an attack can be very high, especially if it is seen as preempting a similar attack on one's own assets.

---

3   The Space Force recently contracted for an AI-based automated space
    warfare adversary to use in war games. This AI tool is likely to perform
    at a much higher level than the humans it is intended to train. "Space
    Force Taps Slingshot to Build AI Adversaries for Orbital Wargames,"
    Breaking Defense, January 15, 2026, https://breakingdefense.com/2026/01/
    space-force-taps-slingshot-to-build-ai-adversaries-for-orbital-wargames/.

OPERATIONAL CONCEPTS

The operational concept for the offensive side of a future war in space is fairly straightforward—build a mix of various means of attack intended to optimize results and employ them together in mutually reinforcing, redundant, and decisive ways. Build in redundancy so if one attack method fails another can still be successful. Acquire enough capacity that after an initial exchange, one can still quickly destroy any attempts at reconstitution. This is not the future of space warfare—it is the present.[4] In this construct, several considerations affect the mix of weapons. For example, weapons that can be effective instantaneously and simultaneously are preferred. Ground- or sea-based kinetic attack mechanisms have the disadvantage of longer flight times and greater difficulty concealing intent. For these surface-launched systems, it only takes a few minutes for low earth orbit (LEO) attacks and a few hours for geostationary equatorial orbit (GEO) attacks. These timeframes aren't prohibitive, but any warning time can be used by an adversary to maneuver away from an attack or take other defensive actions. These weapons do provide better opportunities for damage assessment than some other attack modes, but they also can produce significant space debris. In a major power conflict, this risk might be acceptable, at least to an adversary. The mere existence of potential adversary direct-ascent ASAT inventories today strongly suggests this is the case for Russia and China. I won't attempt to describe an optimal future mix for the United States; it depends on policy choices, operational intent, and threat considerations.[5] There is no real technical constraint on acquiring a desired mix of systems today.

The operational concept for the defensive side of a conflict in space is the asymmetrical dual of the offensive concept. Build an optimized suite of capabilities and use them in close collaboration to preserve and reconstitute the suite of capabilities needed to support terrestrial operations in

---

4   I wrote these words in 2021. They are truer today.

5   One of those policy choices has been to avoid debris-causing lethal mechanisms. The Space Force is achieving this policy goal. I cannot say the same for potential adversaries.

the other domains. The tools and techniques available to ensure the survivability of space-based capabilities have been discussed publicly, albeit speculatively, for many years. They include hardening (shielding against lasers or microwave devices, for example); avoidance (detect the threat and maneuver away); disaggregation (breaking up capability into smaller units—many of which would have to be attacked to defeat the capability); proliferation (just buying more assets to have multiply redundant capability); concealment; replenishment (war reserve assets that could be launched quickly to give a temporary capability as needed); defense (putting sensors and defensive weapons as well as countermeasures on satellites or near them); counter-offense (attack the anti-satellite systems on the ground before they can be used); and deception (for example, make military assets look like much more proliferated nonmilitary systems).

A viable space architecture designed for wartime survivability would include most if not all of these elements, but given my comments on the benefits of surprise, there should be a high premium on concealment and deception. An adversary's well-planned and coordinated attack will not succeed if he doesn't know assets exist or where they are. Even with this feature, a space support architecture would also have to include a sophisticated operational planning and decision support system that could react in near real time to threat actions as soon as information became available. One of the goals of this concept would be to create high uncertainty of attack success in the mind of an adversary. Fear of failure is a powerful deterrent. In addition to increasing the resiliency of space-based systems against attack, it would also be wise to supplement that capability with fallback ground-, sea-, and air-based systems that would provide redundancy and reserve capacity for the most critical military functions currently provided from space to include reconnaissance and targeting; communications; position, navigation, and timing; and attack warning.

Underpinning both space offense and space defense warfare orders of battle, there would have to be a robust space situational awareness capability with the capacity to monitor all activity in space continuously and with high fidelity. A major problem with US space situational awareness today is the degree to which it was designed for a peacetime environment, without adequate concern for resilience or survivability.

Secure, resilient, and responsive access to space is also required. Today, both launch capacity and space systems are acquired on something like a "just in time" management approach. If we are serious about space warfighting and reconstitution, the US will have to acquire a wartime reserve of both launch capacity and satellites.[6] Launch facilities and ground stations will also have to be more survivable and redundant, suggesting either concealed locations or, more likely, mobile systems.[7] C3BM for both offensive and defensive space operations will be highly computationally intensive and highly automated; the operator's role once a serious conflict is initiated may be to simply engage a system of automated responses. An AI-driven course-of-action alternative analyzer with scenario analysis capacity well beyond human reasoning or intuitive potential will be needed to create real-time responses and direct their execution. Human override will be possible and probably insisted upon by operators. Unlike other domains where tactical decisions can and should be decentralized to the "edge," there will be a need in space operations for more centralized decision-making and execution. It's quite possible the space conflict battlespace will be partitioned into geosynchronous, medium, and low earth orbit (GEO, MEO, and LEO) "fights" modeled and directed with some degree of independence from each other. Whether this architectural decision is made or not, the number of simultaneous interrelated actions that will have to be taken would still overwhelm human operators, even with fairly sophisticated tools. Once the space war starts, humans will largely be observers of the actual engagements—if our ability to "see" what is hap-

---

6   The Space Force has demonstrated aggressive timelines to launch satellites on short notice. This will only be useful if both launch systems and payloads are acquired and kept ready for use during a conflict; John Plumb, "Tactically Responsive Space: An Emerging New Deterrence Tool," Center for Strategic and International Studies, October 31, 2024, https://www.csis.org/analysis/tactically-responsive-space-emerging-new-deterrence-tool.

7   The proliferation of commercial launch systems for smaller payloads may partly address this need.

pening has survived. Humans will be involved in decisions about options with longer, less stressing time horizons and more strategic impact, such as when to launch a reconstitution effort or when to unleash a second wave of attacks.

One important thing to note at this point is the Space Force concept conjured up by the description above is not a trivial matter and is far from what we have now. It is also not something that has to wait for new advanced technology to be developed. While the fledgling new service created a few years ago has a very small end-strength and a budget a fraction of the other military services, getting from where we are today to a force that can sustain its functions against a determined adversary and decisively take on that adversary's own space systems will require a substantial investment and a major paradigm shift in how we think about space. During my tenure as the Secretary of the Air Force, we moved forward on integrated C3BM, resilient missile warning, more resilient GPS, and resilient targeting architectures to support the joint force. We also made progress on counter-space systems. At this point, funding is the greatest obstacle to achieving the needed capabilities.[8]

BUILDING BLOCKS

Below, I'll provide a list and some discussion of the specific functional types of systems we will need as building blocks. Some are firmer than others, but all should be analyzed and considered. Many exist today but are not designed for conflict in space. Some are being developed and fielded to meet existing and future operational needs.

---

8   The Trump administration's Golden Dome program overlaps in part with the needs described here. Missile warning and tracking is a common requirement. Space-based interceptors for mid-course ballistic missile engagement may have utility for LEO anti-satellite missions, depending on their orbits and kinematic performance. A space-based boost-phase intercept capability could also be used to engage space launch systems and/or ground-based anti-satellite missiles.

While the functionality of the space-based building blocks is reasonably clear, the optimal combination of mass, orbital planes, individual satellite mission capability, design life, sustainment concept (if any), and combination of defensive features will all have to be the subject of detailed and highly classified technical and operational trade studies. When I originally wrote this chapter, my short answer to how those trades will come out was "I don't know." Many of the trades have been done now, and work to field capabilities has started. There needs to be a significant degree of distributed, disaggregated (not the same thing) constellation designs for survivability, and there will be a premium for concealment and deception in general and in the helpful context of the mega-constellations commercial firms are building. The traditional NRO concepts of large, expensive, and highly targetable satellites will not meet the warfighting needs of the DOD. During my tenure as the Secretary of the Air Force, I had a great working relationship with the NRO led by Director Christopher Scolese and his deputy Troy Meink, who became my successor as Secretary of the Air Force. These impressive leaders shared the vision of future warfighting systems that would support both operational and intelligence needs, and we were able to initiate programs designed to directly support the joint force against a peer competitor, particularly in moving target detection and tracking.

There is, however, a probability any military satellite constellation will not be adequately survivable in space against a determined adversary. If that is the case, commercial systems in existence and in development offer the potential for backup imaging and communication capabilities that can be called upon in wartime. Reconstitutable space support systems launched in conjunction with an operational need, and which are not expected to survive beyond a short operational window, may also be needed as a building block—together with a survivable mobile launch capability. The Space Force is actively building relationships with commercial space firms and experimenting with rapid reconstitution. Beyond that, there is a lot of engineering and operational analysis still to be done. China, and to a lesser extent Russia, will respond to anything the US fields, and we must take responsive threats adequately into account. That should happen before we make major investment commitments.

Our future offensive mix of capabilities should be quite broad. China is proliferating operational space assets using hundreds of satellites, and we need cost-effective responses with deep magazines, low cost per kill, and high engagement rates. Our capabilities should include the ability to attack relevant space-related assets with conventional and unconventional weapons. Directed energy in the form of ground-based lasers, in particular, should be considered for inclusion. Soft-kill mechanisms, including EW and cyber, should be fielded as well, and continuously upgraded.

Here's the mix of systems we will need or should at least consider.

Offensive Counter-Space Systems

Direct-Ascent ASAT: Dedicated ASAT systems have been fielded already, but not by the US. The US did test an air-launched ASAT in the '80s but abandoned the program.[9] More recently, a US Navy Standard Missile was used to intercept a deorbiting satellite as a safety precaution, but this missile has very limited ASAT capability because of its kinematic characteristics. Russia, China, and others have conducted direct-ascent ASAT tests very recently. A ship-based or ground-based system would be more responsive and easier to keep in a high readiness state than an air-launched system. It could also be dual-use and affordable in reasonable numbers. Debris limitation is a design criteria. Both LEO- and GEO-capable systems should be considered. The very high numbers of fielded and projected adversary LEO ISR satellites argue against this option as the primary LEO engagement mechanism, however.

Orbital ASAT: Anti-satellite systems of various types could be placed in orbit in anticipation of a conflict. There are press reports that Russia

---

9   Overseeing this anti-satellite program that had been successfully tested in 1985 was part of my first job in the Pentagon starting in 1986. I was unable to convince the Congress to authorize continued testing until or unless the Soviets conducted more tests of their system. They had completed testing and the Soviet system was already in operation. As a result, we couldn't test and canceled the program for that reason.

intends to place nuclear weapons in orbit as anti-satellite weapons.[10] If true, this would be highly destabilizing and a violation of the Outer Space Treaty. What I have in mind here is conventional or at least nonnuclear lethal mechanisms.

Directed-Energy Weapons: Over several decades, the technology for ground-based laser anti-satellite systems has advanced significantly, but never reached levels where an effective weapon could be fielded, particularly if efforts to harden against laser energy are taken into account. The combination of brightness and atmospheric correction needed have been too much to overcome, but given the advances to date, I believe this capability could be fielded in the timeframe of interest. A laser ASAT will likely operate from a fixed location and represents a high-value target to an adversary. Mobile systems are possible; that feature would dramatically increase survivability. An advantage of a laser ASAT is that it can conduct many engagements over time, particularly against LEO targets. This feature makes such a weapon potentially very cost-effective. Orbital directed-energy weapons are also a possibility, but it would be challenging to ensure their survivability.

Military Support Systems

We already have systems in space to support other domains. These are all enduring needs.

Intelligence Collection, Analysis, and Dissemination: Electronic Intelligence (ELINT) is composed of Communications Intelligence (COMINT) and Signals Intelligence (SIGINT), Imagery Intelligence (IMINT), and Measurement and Signature Intelligence (MASINT).[11] In addition to peacetime intelligence collection, ELINT, IMINT, and MASINT systems

---

10   Unshin Lee Harpley, "DOD Official Confirms Russia Is Developing an 'Indiscriminate' Space Nuke," *Air & Space Forces Magazine*, May 2, 2024, https://www.airandspaceforces.com/dod-official-russia-indiscriminate-space-nuke/.

11   ELINT encompasses both the communications content and noncommunications signals from an adversary source. MASINT refers to a range of nonelectronic phenomena.

can provide tactical targeting support to ongoing operations and information critical to EW systems. A future space domain warfighting construct would include survivable versions of these systems and, equally important, the ability to process the information they collect into operationally relevant data and information at operationally valuable speeds. Some collection will also be done with terrestrially based and airborne systems. All these systems should have the capacity to promptly go from raw intelligence to battlefield situational awareness as to geolocation and identity. They should also support prompt countermeasure development with a high degree of automation. Data analytics and machine learning are important components of these future capabilities. Perfection isn't achievable, so sensors and processing should be designed for the highest payoff practically achievable operational benefit, with increasing performance over time. Current capabilities are far from the achievable goal.

Land, Sea, and Air Targeting and Engagement Support: While space-based sensors have been used to collect intelligence on fixed targets since the early days of the Cold War, the ability to detect, identify, and track moving targets has been more limited. In the future, if they can be made adequately survivable, space-based sensors, supported by artificial intelligence and secure communications, will directly support operational targeting and engagement. Both the US and China are pursuing this goal aggressively. The US is committed to defeating aggression, whether it comes primarily by land or primarily by sea. This is best achieved by destroying moving targets, on the sea, in the air, and on the ground. The combination of long-range precision weapons and precision tracking of moving targets from space with position update as needed to support engagements is the key to defeating aggression in the future. Achieving this goal is dependent on more than technology, however, at least for the US. To some degree, the US is burdened by a historical division of labor between the intelligence community and the defense community that had its origins in the 1950s. Technology is eviscerating the line between intelligence information needing processing and analysis before it is useful to military operational decision-makers and information that can be acted upon immediately without the need for an intermediary. For airborne sensors like AWACS and JSTARS, this seam was already eliminated, but the same is not true for space-based sensors, which have traditionally been under the control of the intelligence community and not the operational community. We made progress on resolving those differences during my tenure as the Secretary of the Air Force, but those cultural and organizational seams are still an obstacle to obtaining the operational capabilities we will need to succeed.

Missile Warning and Tracking: Operationally useful missile warning must be survivable and made available in near real time for warning, defense systems, and counterstrike decisions. Threat systems of interest include long-, intermediate-, and short-range cruise and ballistic missiles, including hypersonic missiles. Tracking from origin and during flight will be needed to support defense system engagements on compressed timelines. During my tenure as the Secretary of the Air Force, we initiated programs to provide resilient distributed missile warning and tracking. Those systems are in the process of being fielded. Under the Trump administration they are likely to be folded into the Golden Dome concept, but they have high utility independent of the Golden Dome goal of a robust national defense. Strategic deterrence depends on survivable and highly reliable missile warning as does regional air defense. Similar functionality for longer-range air-to-air missiles is highly desirable, if achievable.

Position, Navigation, and Timing: As mentioned elsewhere, this functionality has become ubiquitous and is essential for current systems with a wide range of functionality. The timing service provided by GPS is essential throughout the military services, and the current constellation in MEO is more vulnerable to attack than most realize. The latest improvement, GPS III with the M-code waveform, may not be adequate against advanced threats in the future. While providing this service from orbit is by far the most cost-effective approach in peacetime, it creates serious operational risk in wartime. Future space domain concepts should include this service, but acceptable and more survivable alternatives and provisions for graceful degradation if space-based PNT is lost will also be necessary.

COMPLEXITIES

Fundamental Space Domain Operational Needs

Actionable Timely Space Asset Attack Warning: The military uses something called indications and warning metrics to project the likelihood of an attack and detect situations in which that likelihood is increasing. For conventional operations, these include logistics movements, exercises that could conceal mobilization or attack preparation, and key leader behaviors. We will need a combination of intelligence indicators, space and ground activity monitors, and artificial intelligence-based data analytics to measure and assess the space operational environment continuously. This is an obvious application of some forms of advanced data analytics and inference machines, but it has to be well supervised by humans, and spoofing or deception is a major concern. This type of system can be used to decrease the instability discussed above, but it entails risk. Mutual confidence-building measures may be part of this construct, if they can be negotiated and verified.

Command, Control, Communications, and Battle Management: During the Obama administration, we moved forward with the creation of a joint space operational center, now known as the National Space Defense Center (NSDC). The NSDC is intended to coordinate defense, intelligence, and commercial space activities. The more recently formed Space Development Agency (SDA), which was successfully integrated into the Department of the Air Force during my tenure as the Secretary of the Air Force, is fielding a "transport layer" constellation to provide communication services to national security space assets and terrestrial users. We need truly integrated and survivable C3BM for space assets under a single unified wartime commander, presumably US Space Command. We will need communication networks adequate to support both our space-based services needs for data movement and our communication needs for space situation awareness, as well as control of our space assets, and conducting offensive and defensive operations in space. This system should be designed for graceful degradation under attack, very high levels of cybersecurity, and with as much preprocessing of data as practical in orbit to reduce bandwidth requirements. The C3BM system should also have highly automated operational analysis tools for assessing situations, evaluating options, and recommending or implementing courses of action.

Launch and Ground Control: Our ground control sites are generally fixed and have significant operational footprints, making them subject to adversary targeting. In the future, we will need to acquire launch systems that can be resilient, probably through mobility and concealment, and tailored to the classes of payloads (small) and orbits (varied) we intend to utilize for national security missions. We will need an inventory consistent with our expected wartime needs. Ground control systems will need to be similarly robust, redundant, and survivable. Commercial constellation control has advanced dramatically and is now highly automated, even for large constellations. This technology is maturing quickly.

Logistical Support: There are nascent efforts ongoing to provide on-orbit logistical support to satellites. It's an open question whether this makes sense for many warfighting assets. It may be cost-effective to refuel, repair, or augment satellites already on orbit in peacetime. But, if we plan to conceal military assets in space, visibly servicing them is problematic. If we expect our satellites to have short operational lives because of threats, then acquiring the capability to service them in orbit may not be cost-effective. During my tenure, we continued to conduct experiments in this area. I'm open to the possibilities in this regard, but not certain of how the trade-offs are likely to come out as our warfighting space order of battle becomes more mature.[12]

---

12 In one area, however, the Space Force is moving forward with a requirement
    for on-orbit refueling. Theresa Hitchens, "In a First, Space Force to Require
    Refueling Capability for Next-Gen Neighborhood Watch Sats," *Breaking Defense*,
    September 24, 2025, https://breakingdefense.com/2025/09/in-shift-space-force-
    to-require-refueling-capability-for-next-gen-neighborhood-watch-sats.

Multi-Domain Considerations: Space supports all the other domains, and the existence of support from space or the lack of it will be decisive in all other domains. For air, land, and sea surface operations, space systems provide critical situation awareness, warning, and targeting information. Space can also provide secure directional communications needed to support operations in all other domains. Even for undersea systems, connectivity from the sea surface onward will be through space systems. Today, we also have a strong dependency on space for PNT for the other domains. While positioning and navigation can often be provided to acceptable levels for many purposes by other means, our dependency on space for high-fidelity time references and synchronization is ubiquitous and essential in all domains except subsea, at least for current systems. Space operations are, of course, dependent on land, sea, and airborne C3BM nodes and launch facilities as discussed above. I don't foresee conventional weapon effects being generated from space anytime in the foreseeable future (the absentee ratio and launch costs, even though they are shrinking, are prohibitive).[13] The US military services are all still wrestling with how to define JADC2 and Multi-Domain Operations (MDOs)—ubiquitous buzzwords for the last several years. I'll have more to say about this in the next two chapters. In any event, the concepts I have described in earlier chapters are very dependent on these space-based functions and associated C3BM.

Space Surveillance and Situational Awareness: As space has become more congested and the mix of debris and small satellites has increased, it has become increasingly difficult to track and monitor all objects in orbit. Throw into this the potential for threat intelligence collection, hard or soft interference by adversaries, and deceptive objects deployed for a variety of reasons. The range of space surveillance needs spans continuous launch monitoring, space object tracking, awareness of maneuvering in space, and any proximity operations. Those are peacetime needs. In wartime,

---

13   The absentee ratio refers to the fact that LEO systems circle the earth about every ninety minutes, which means ten to twenty satellites would need to be deployed to ensure there was always one space-based weapon over a target continuously.

add attack detection and characterization and damage assessment and the requirement for survivability.[14]

Strategic and Tactical Reserves: By strategic reserves, I mean those needed for a longer conflict. By tactical reserves, I mean those needed immediately to support the initial stages of a conflict. Reserves can be terrestrial or on orbit, but they must be survivable in either case. This is as much a question of when reserves would be released as it is a distinction about the nature of the payloads and launch vehicles or weapons. Reserve considerations apply to both space defense and offense systems.

Tactical Mission Variations: The future concepts I've described attempt to deal with a scenario in which an all-out surprise attack on our space assets is possible and in which we seek the capacity to launch such an attack on our adversary. One can imagine scenarios in which something less is needed or desired. Three examples come to mind. First are steps to control escalation with actions short of physical destruction, such as cyberattack or EW attack on specific assets. Second is a destructive surgical attack on a specific unacceptable threat capability, such as a space-based ASAT directed-energy weapon or a nuclear ASAT deployed in space. Third is a defensive operation to defeat a potential set of attackers in the process of moving into position for a future attack. In these and other situations, it would be highly desirable to have a range of options available for national decision-makers.

Target Identification/Collateral Damage Avoidance: The biggest operational concern from a cost-effectiveness point of view is to use the inevitably limited number of counter-space weapons in the US inventory to attack actual targets as opposed to decoys or nonthreatening civil and commercial systems. Given the potential opportunities to conceal or disguise threatening military systems in space and the advantages that might

---

14   Traditionally, the Air Force, and now the Space Force, has provided peacetime space object cataloging, tracking, and collision-avoidance support. That responsibility is transferring to the Department of Commerce, but the Space Force will have a continuing requirement to monitor and predict space objects in general for operational reasons.

provide, this is a real concern. It drives the need for high-fidelity intelligence collection, including on-orbit collection. The collateral damage problem in space is generally not civilian populations, but damage to civil or private space systems and to assets belonging to neutral parties.[15] This can happen through intended but mistaken engagements of targets or through unintended collateral damage from debris or imprecise lethal and nonlethal mechanisms. In a great-power conflict, that might be viewed as existential; I doubt these concerns would constrain our adversaries, and they might not constrain the US for very long, but they should be taken into account in the design of the future space concept.

Design Requirements and Considerations

Commercial and Third-Party Space Considerations: Many nations and private companies operate space-based systems, especially large LEO constellations. There are a host of policy issues associated with the prospect of high-intensity military operations in this environment. A future space warfare concept has to consider the potential for threat systems to hide among commercial or third-party constellations. It must take into account the potential for an adversary to rely on military functionality (intelligence collection, target identification and tracking, communication, PNT) provided by neutral nation-states or through national or globalized corporations. This is as much or more a diplomatic problem or a policy problem as a military design issue, but if the diplomatic and policy problem is not solved, the military problem will be much more challenging.

Confidence in Automated Behaviors: Most behaviors in space are already automated, once a command is given, but that's for local short-term behaviors. The big problem is the lack of time for human involvement in the many operational decisions that would have to be made on highly stressing timelines to fight a determined adversary trying to exe-

---

15   Civilians could be implicated in attacks on ground-based space infrastructure. There will be some civilian activity in space itself, but I don't foresee a significant human presence in space in the timeframe of interest here.

cute a well-planned coherent attack operation against our assets in space. Selecting a generic option or class of options and enabling automated execution would be all that was possible for humans to achieve, at least without high risk of failure. I am imagining a Pearl Harbor-style attack, combining decisive scale and surprise, because that's the best option for an adversary. Why would an adversary do anything else? The only way to gain confidence in the automated operational response we would need is through extensive simulation and war-gaming. Elements of a response and local behaviors (maneuver or deploying countermeasures, for example) could be verified on orbit as well as in hardware and software in-the-loop simulations, but a large-scale response could only be verified in simulation.

Constellation Management: The task of keeping a large complex constellation of satellites functioning as a unit is nontrivial. The constellations that are being fielded by the SDA will need continuous management, as will the proliferated commercial constellations currently being fielded. For peacetime operations, this problem has largely been solved, but the function becomes much more challenging when the constellation is under attack, and some subset of its assets is destroyed or disabled. This function can be automated; in fact, to a large degree, it already has been with large constellations like Starlink. However, the automation for war-fighting situations will need to cope with a wide variety of conditions an enemy may try to impose.[16]

Countermeasures: There are a range of defensive features that should be considered in each space system design. These include maneuver, decoys, self-defense, signature management, and hardening. As satellites increase in mass, the ability to include some countermeasures, such as maneuver, effectively tends to decrease.

Cyber and EW: These are absolutely an important part of the equation, with major design implications for both offensive and defensive space systems. All satellites depend on software and hardware subject to

---

16   An example is an attempt to blow a temporary "hole" in a missile warning
      constellation so that an attack can proceed without being detected or characterized.

attack at any time in the product life cycle, including in orbit. For any satellite system, there is likely to be a range of potential target portals, including all RF antennas. Satellites depend on radio frequency or laser communications and a variety of sensors for mission performance and command and control. Ground systems can also be a critical node and a single point of failure in an EW or cyberattack.[17]

Debris Generation and Avoidance: There is military value in being sure a targeted system is actually destroyed. Kinetic systems with highly visible and conclusive damage mechanisms provide this, but in space, a high-speed collision or explosive weapon is likely to also create persistent and growing debris fields that threaten everything in their orbital path. Soft kill can be very cost-effective, but may leave unanswered questions about the actual status of the target. There is no perfect solution here, but when possible, debris generation should be avoided—even for hard-kill systems. For some systems, debris sensing and avoidance may be feasible and cost-effective, but I have not seen any designs that attempt to achieve this and I'm skeptical it can be accomplished. A sophisticated Space Situational Awareness System can provide warning for larger items, but in an active conflict is likely to be overwhelmed.

Default Behaviors—Loss of Contact/Control: All satellites have provisions for default behaviors in various circumstances. Generally, they are designed to preserve functionality and create opportunities to correct any onboard problems. This is a little more complex for warfighting systems that may be under attack as opposed to experiencing a malfunction, but the design considerations are similar. Depending on the operational profile for the satellite and system and the expected reliable design life, some degree of redundancy and enhanced resilience is likely to be cost-effective for warfighting systems.

Energy: I don't see a need for novel energy sources for the space domain, but any technology that matures and has operational, economic, or envi-

---

17   One of my biggest headaches in government was the ground control
      software program for GPS III, a system called OCX. The driving technical
      issue causing extensive cost and schedule overruns was cybersecurity.

ronmental benefits would be adapted. Several technologies for propulsion and/or energy generation and storage are in some stage of development. As discussed earlier, concepts like on-orbit refueling could also be utilized if proven to be cost-effective.

Escalation Control: While I believe we should take a large-scale attack seriously, one can argue that less decisive attacks in space could be used as part of an escalation-control ladder meant to deter the United States. Attacks that do no permanent or visible damage and are reversible are particularly attractive for this purpose. This should be a consideration in our offensive and defensive concepts.

Ground Station Resilience: Ground stations or terminals in general can be an Achilles' heel for space systems. Especially for forward-located militarily important ground control systems, there should be mobile survivable backups in addition to any fixed facilities. Hardening and/or concealment and deception should also be considered.

Hardening Options: Physical hardening of space assets against kinetic attack isn't feasible, but some hardening can be designed for other weapons. Hardening can be achieved against laser attacks up to a certain level through mechanisms like shutter control for optics or reflective nonabsorptive materials. Hardening against high-power microwave and electromagnetic pulse or radiation (both generated by nuclear detonations) can also be accomplished, but at a significant cost.

Humans in the Concept: I'm not projecting any need for human warfighters on orbit. On the surface or possibly in the air, they will be needed for command and control and for key decisions about implementing automated operational courses of action. Because of the operational importance of space, I would expect specialists in space systems and operations to be deployed in the most forward C3BM elements that oversee joint and combined operations, as well as at headquarters dedicated to commanding and controlling space operations overall. The Space Force has been building Space Component Commands with the various operational Combatant Commands, but there is still much to be done to define the roles of these organizations and to give them the tools they will need to be effective.

Identification Friend or Foe: Space is big, and movement is constrained by orbital mechanics. Most objects, especially friendly objects, are known to their operators and controlling national organizations—for the US, it's the US Space Command. At first blush, I'd guess IFF may not be needed or cost-effective because of the infrequency of encounters between a friendly and an unknown but in fact friendly or neutral satellite. Militarily, IFF systems for aircraft are primarily there to prevent mistaken engagement of friendly or neutral aircraft and for air traffic control. Traditionally, they involve active interrogation of a potential threat and require a cooperative response system on any nonhostile interrogated platform.[18] Depending on what mix of military and commercial systems, self-protection measures, and the mix of tactics and threats we expect to field and encounter, it's conceivable the ability to interrogate or self-identify to an approaching object in space would have some value, but I don't see it as a priority. With international agreement, some form of IFF could be a confidence building measure for any close proximity operations involving multiple nations' satellites.

Interoperability: The US has and will have a suite of space systems with enormous military value to the US, the joint force, and our partners. There are interoperability needs associated with integration of airborne, terrestrial, and space-based systems, as well as with allied systems and commercial systems. The barriers to interoperability are much more about management, cost, and organizational incentives and culture than technology. Internal to the US, the biggest interoperability issue within the space domain, as indicated earlier, is the separation between intelligence and military space assets. The concepts I've laid out are very dependent on the efficiencies associated with space-based targeting and engagement

---

18   IFF exists for aircraft because airspace can be crowded even in wartime, and Air Force pilots don't like the idea of being shot down by an Army air defense system (such as the one I operated in West Germany in the '70s).

support, assured communication, and PNT support; interoperability for those tasks isn't optional or nice to have—it is essential.

Launch Services: The last decade or so has seen some dramatic reductions in launch costs, notably with reusable boosters. The next major advance, which I believe will be successful, is the very-large-scale reusable systems, particularly Starship being developed by SpaceX. Current military launch services are provided by commercial firms and span very small payload capacity to LEO up to very large payload capacity to GEO and beyond.[19] They also provide efficient ways to move multiple smaller payloads into LEO and other orbits. The concept for the space domain requires the additional support of survivable and responsive launch services for payloads for reconstitution of disabled or destroyed orbital support systems. I expect this to mean road-mobile and/or aircraft-launched systems capable of nominally putting up to one- or two-ton satellites into LEO. The commercial launch community, currently well populated, can supply some of these needs, but not all. Some wartime capability may have to be provided by military-operated systems. Large-scale fixed-launch locations used for national security launches in the continental US may well be targeted with long-range conventional weapons.

Legal Constraints: The major constraints today are the bans on weapons of mass destruction in space and the possible legal liabilities of creating large debris fields that can damage third-party assets. Neither of these is an effective constraint on the concept I've described. There are also legal treaty limits under the Outer Space Treaty of 1967 on militarization of locations like the moon. The practical extent of those limitations is currently being explored in the courts, but they probably don't apply to the concepts described here. There is some international and domestic policy interest in constraining weapons in space and ASAT systems, but I don't

---

19 In 2014, when I was the Under Secretary for Acquisition, Technology and Logistics, then-Lieutenant General Ellen Pawlikowski and I led the transition from DOD buying rockets for space launch to contracting for launch services provided by commercial firms. That approach was successful and has continued. How this approach would function in wartime, if it could, needs to be addressed.

see it gaining real traction in the form of international agreements. There is also a fairly widespread interest in defining acceptable norms of behavior in space, with or without legally binding agreements. These might include prohibitions on proximity operations and rules about deconfliction and debris creation.

Mission Planning, Including Dynamic Replanning: Planning space military operations, largely a mix of satellite attacks and defensive measures, will be a complex endeavor, even before the enemy makes its first move. Once both sides are actively conducting combat operations, it will become even more important to have developed an automated and sophisticated course-of-action engine that can create and evaluate multiple integrated options in an operationally viable time scale. If we can survive an enemy's first move in space, US space forces must adapt a dynamic replanning-at-scale approach as well.[20]

On-Orbit Servicing: This should be considered for inclusion in the concept, but I'm not 100 percent sold on it at this point, at least for military satellites in wartime. On-orbit servicing pays for itself if the extended life and functionality of the satellite is worth the cost of the service. In April 2025, the Space Force announced that it is proceeding to include on-orbit refueling for its next on-orbit space surveillance system.[21] The current commercial trend, however, is toward low-cost, relatively short-lived systems. If on-orbit servicing makes sense on a cost basis in peacetime, then it should be included for that reason, but servicing during a conflict involves some additional trades that don't obviously support the concept. However, once on-orbit servicing buys its way onto the concept for peacetime, the delta cost for a warfighting capability should be small. Servicing can take several forms, but it's often refueling, replacement of

---

20   An overused and ill-defined phrase in US military circles right now is "the speed of relevance." This is one of many use cases where unassisted human decision-making is incapable of acting effectively at that speed.

21   Theresa Hitchens, "In a First, Space Force to Require Refueling Capability for Next-Gen Neighborhood Watch Sats," *Breaking Defense*, September 24, 2025, https://breakingdefense.com/2025/09/in-shift-space-force-to-require-refueling-capability-for-next-gen-neighborhood-watch-sats/.

a failed module, or movement to a new orbit. I doubt that repair of battle-damaged satellites would be cost-effective, but one can imagine steps like replacing damaged solar arrays as a possibility. If the cost of servicing exceeds the cost of a new satellite, then the answer is pretty clear.

Operational Planning and Rehearsal: Our capacity for planning and rehearsal in the future will become much more sophisticated and high fidelity. Some forms of AI will play a major role as we are forced to rely on simulation and autonomy for planning. Rehearsals will be virtual, of course, but as we get into actual operations, it will be important to learn from real-life experience immediately and feed that data into machine learning contexts to improve planning.

Operationally Responsive Space: This concept has been around for a long time. It has received support from the Congress to keep it alive over the years, but very little within the Air Force or Space Force until recently. Recent tests like the Victus Nox test in 2023 have demonstrated the viability of short-notice launches, but we have much further to go down this path.[22] Given the current and projected threats to our space assets and ground-based infrastructure, we may well need operationally responsive space in the forms of responsive launch, war reserve payloads, and payloads that can be activated as soon as they are on orbit. All of this is useful to support the concept I have described above and to make it more operationally resilient, but there are many other shortfalls that need to be addressed that are higher priority at this time.

Physical Security: A future space warfighting force will need adequate physical security for terrestrial systems. The US will also need physical security for on-orbit assets, but it's an open question of how much and in what form. The threat I'm concerned about for our on-orbit systems is peacetime preparation of the battlefield activities by potential adversaries. They will be motivated to conduct proximity operations to gather

---

22   SSC Space Safari, "US Space Force Successfully Concludes VICTUS NOX Tactically Responsive Space Mission," Space Systems Command, February 19, 2024, https://www.ssc.spaceforce.mil/Newsroom/Article-Display/Article/3679056/ us-space-force-successfully-concludes-victus-nox-tactically-responsive-space-mi.

intelligence, covertly damage our systems, or implant malicious devices (digital or physical) on our assets for future use. Protecting against these threats should be a design requirement for future systems. Both passive and active security measures should be considered, and an enforceable declaratory policy about threat proximity operations should be part of the package of security measures.

Reliability: The cost of US military space systems has been driven by stringent reliability requirements. The cost of the spacecraft themselves, the cost of launch, and the lead times associated with new space systems have all argued for very high reliability in the past. As launch costs have come down and commercial payloads have proliferated, this equation has changed dramatically. Designing for reliability in space is also much better understood now than it once was. The potential for attrition and even high loss rates in wartime is also an argument for reduced reliability requirements. My sense is we will still want high initial reliability, especially for wartime reserves, but less required lifetime in orbit.

Requirements Creep: Conscious attention should be placed on avoiding cost imposing marginal or low return-on-investment requirements. The natural tendency in all domains, including space, is to add more requirements to the design until the concept crashes from its own weight. A lot of the items on this list are good examples of potential requirements creep if not managed carefully. The whole point of all the concepts is improved cost-exchange ratios over current systems. In space, we have an environment where the previous assumption of impunity and the small numbers of targets we were concerned about permitted us to ignore cost-exchange ratios. That assumption is no longer valid.

Responsive Threats: There will be an action-and-response paradigm in space as there is in other domains. As the US moves to match and exceed what our potential adversaries are doing, it can expect first increasing numbers of offensive systems, followed over time by more sophisticated and lower-cost-per-kill ASAT concepts, across the suite of options available. On the defensive side, the US should see efforts to conceal space-based capabilities, use of third-party commercial alternatives globally to support military operations and complicate US planning, and continued use of airborne and terrestrial alternatives to space, including high-altitude long-endurance UAVs. China is increasingly relying on space to support military operations, especially targeting to support long-range precision weapons of various types. For Russia, a mutual denial of space-based assets could provide benefits relative to the US and might be a more than acceptable result. Press reports of Russia potentially placing nuclear weapons in space to indiscriminately destroy large numbers of satellites are consistent with this perspective.[23]

Secure Communications: This is a necessity for space-based operations and for support to other theaters from space. I expect continuing advances in the relevant technologies, both RF and laser-based. Quantum-based phenomena are starting to emerge, offering enhanced security from both interception and decoding as well as very high data rates. The SDA has been leading the effort to develop and deploy distributed systems that meet these objectives. Significant progress was made toward this goal while I was the Secretary of the Air Force under the leadership of Derek Tournear at SDA.[24]

Single Points of Failure: For space in particular, our adversaries will actively search for an Achilles' heel in the system. Any vulnerability in the C3 system is a potentially catastrophic problem. Ground control system nodes are a particular concern. Given the importance of space situational awareness sensors for threat detection and identification, reliance on a vulnerable set of ground and space-based assets for that function could be problematic. The idea of warfighting in space is so new and unique, and its importance is so high, extensive independent red teaming to detect and help eliminate cheap catastrophic attack options is a necessity.

---

23   Marc Berkowitz and Chris Williams, "Russia's Space-Based, Nuclear-Armed Anti-Satellite Weapon: Implications and Response Options," National Security Space Association, May 16, 2024, https://nssaspace. org/wp-content/uploads/2024/05/Russian-Nuclear-ASAT.pdf.

24   Courtney Albon, "Space Development Agency Launches First Operational Satellites," *Defense News*, September 10, 2025, https://www.defensenews.com/ space/2025/09/10/space-development-agency-launches-first-operational-satellites/.

Training, Experimentation, and Testing: Many aspects of warfare in the future will be unprecedented and vastly different from our historical experience. This may be most true regarding conflict in space. The US and its adversaries will all have to rely on the training, experimentation, and testing that can be conducted through simulation anchored in a limited number of experiments. This will put a high premium on accurate modeling and simulation. Equally important it will place a high premium on the willingness of senior military leaders to accept what the models and simulations are saying about the likely outcomes of future conflicts. My experience has been a general reluctance by senior military leaders to except modeling and simulation results that don't conform to institutional preferences and intuition. This tendency could prove very dangerous, especially when applied to the decisive domain I expect space to be.

Weather Implications: The main design concern here is the well-known problem of space weather, especially the environments associated with high sunspot activity levels. It's a fairly predictable phenomena that space systems engineers are well familiar with. One would like to avoid a predictable and potentially exploitable transient vulnerability brought on by space weather. Terrestrial weather is also a factor in requirements for almost everything in the concept I have described. It drives some sensor requirements and affects launch systems and ASAT systems, for example.

Other Needed Military Functions

Disaster Relief and Humanitarian Assistance Support: Military space systems can provide a wealth of information in support of disaster relief and humanitarian assistance. There are and will increasingly be several civil earth monitoring systems fielded by various states and agencies, so the role of military systems will be to support and augment those in existence for civil purposes. There is a balance to be struck between revealing military space surveillance capabilities and supporting civil functions like disaster relief. This balance should be considered during architecture and constellation design when marginal changes that would be problematic later can be incorporated relatively inexpensively.

Intelligence Support and Support to Intelligence: For the US, there may be a future division of labor between US intelligence and US military space that is going to result in separate systems operated by each to collect similar information for different purposes, on different time scales, and with different degrees of fidelity and resilience. This division is driven by the fact that the military needs less precise but more timely and survivable operational support from space, while the intelligence community needs exquisite quality products for peacetime national intelligence purposes. While we had impunity to operate in space, intelligence and military needs could be satisfied with the same space systems, at least up to a point. For decades, the military has depended on the intelligence community to field certain types of space systems, often co-funding systems with mixes of National Intelligence Program and Military Intelligence Program dollars budgeted by each community. It was discovered decades ago that very sophisticated national intelligence assets could not connect to and support operational needs in a timely manner, if at all. As a result, there has been some success over the years in ensuring intelligence systems procured and operated by the NRO or others could provide operationally useful data to military users in theater. This was acceptable until the advent of robust adversary anti-satellite inventories. The US has the choice to make the intelligence community's space systems survivable enough for wartime use, and possibly reduce the fidelity of information peacetime users demand, or to create separate peacetime collection and warfighting space ISR systems. At this point, I'm of the opinion that in some cases at least, separate architectures will be needed, but that decision has not been taken and will be resisted by both sides as long as possible because of the cost and/or potential loss of resources it entails.

Whether this occurs or not, there is an enduring need for both communities to support each other, in peacetime and in wartime. That connectivity and cooperation should be designed into any future space architectures. During my tenure as the Secretary of the Air Force, we worked closely with the NRO on space-based moving-target sensing systems that could feed operationally relevant information directly to military users. We encountered strong resistance, however, from the National Geospatial Intelligence Agency, which traditionally processes data provided by the NRO sensing systems before it is passed to operational users. If we are going to achieve the timeliness and responsiveness needed to implement the concept described here, those bureaucratic obstacles must be overcome.

Missile Warning—Regional: The US has alliances and close relationships with many nation-states that are threatened by missile systems operated by hostile neighbors as well as forces deployed forward around the world. These threats are only going to grow in severity in range, quantity, accuracy, and sophistication.[25] Some of the nation-states threatening our allies already have nuclear weapons and others may in the future. Our space systems should have the capacity to provide timely direct warning and attack assessment, and possibly engagement support, to global partner countries most concerned about these regional threats as well as to our regionally deployed forces.

Position, Navigation, and Timing: I discussed PNT above as a core military capability, but it's much more than just that. It's impossible to overstate how ubiquitous and important PNT services are to both the military and civil communities. GPS and its several global clones are used throughout the world and in myriad applications, many of which are not generally recognized. The US military will undoubtedly continue to provide this global service in peacetime indefinitely and will continue to modernize PNT services, including adopting more security against threats like jamming and cyberattack. PNT will also continue to be a lucrative target

---

25 Israel is a good example. I've been in Tel Aviv for rocket attacks and once spent time in a hard room with the Israeli Minister of Defense during an alert. Israel's defense systems like the Iron Dome have coped adequately with current threats because the attacking rockets are unguided and only a small percentage have to be intercepted to protect populated areas. Once Israel's enemies acquire precision missiles, the equation will change completely. The conflict between Iran and the US and Israel is providing evidence that this has already occurred.

for our adversaries and potentially an adversary vulnerability the US can exploit for its own purposes. I don't disagree with the need to take steps to protect GPS, but I believe the US should aggressively pursue alternatives and backup capabilities to GPS for military systems in general and possibly for civilian use. We can't afford to be so completely dependent on one system for such crucial services.

Special Operations Support: Military space systems should be integrated with other standoff systems that support special operations, such as counterterrorism and counter-proliferation. The last two decades of emphasis on counterterrorism have led to tight integration of relevant national capabilities, and this should continue for the full range of future special operations applications as new systems more focused on great-power competition are fielded.

Weather Sensing and Forecasting: Commercial and civil space-based weather monitoring and predicting systems have become more sophisticated and detailed. As this has happened, the demand for military-specific space-based weather systems has diminished to almost zero.[26] The trend to ever more precise civil weather forecasting and monitoring is certainly going to continue, especially as more severe weather events occur. The problem for the military will be the availability of this information during great-power competition. Future architecture designs should consider the potential that at least some of the national and internationally sourced weather information may not be available during a conflict.

CONCLUSION

Unlike the other domains, the United States does not already have a warfighting force fully prepared to contest control of the space domain. We are in the process of building that force. As the Secretary of the Air Force,

---

26   There are some unique military needs, such as more precise sea surface winds, better cloud cover predication, and data related to local off-road ground mobility. However, in tight defense budgets, there hasn't been much willingness to pay for systems to obtain this information.

I often used an analogy to describe our space domain situation: The US is like a country that has a merchant marine and wakes up one day to the realization that it needs a Navy. The warfighting assets that we need for conflict in space are a complete departure from what we have traditionally fielded. This is starting to change, and a lot of progress has been made over the past few years, but even more than in other domains, we need a transformation in military space, and we need it as quickly as we can get it.

CHAPTER 7

WARFARE DOMAINS– CYBERSPACE

I decided to categorize cyberspace as a separate domain. I believe we need to think about cyberspace similarly to the more traditional domains of warfare. The traditional domains are geospatially distinct, and each has its own environmental characteristics that are determinative of how conflicts occur in those domains. All the domains I'm describing are related to one another and interdependent. Warfare doesn't respect human definitions of boundaries—it's a very pragmatic endeavor. Like EW, cyber operations will take place in all the geospatial domains. That aspect of cyber, like EW itself (which I am not treating as a separate domain), must be dealt with in the context of conflict in the geospatial operational domains.[1] The operational cyberthreat dictates cybersecurity be a high priority in every domain and throughout the whole spectrum of operations and support functions. It also provides some potential opportunity for decisive defeat of an adversary through military operational cyberattacks. It is a given to

---

1   There is an argument that the struggle for control of the electromagnetic spectrum should be treated as a domain. I didn't take that approach because I don't see the EW struggle as separate from warfare more generally in the physical domains I have discussed. I'm focused on cyber in this section as a place of conflict independent of any other domains.

me that future military systems across the board must be designed to be cybersecure and resilient to cyberattack. The US also has to do its best to prepare and field cyber offensive weapons targeted at adversary military systems and functions. I've alluded to this in previous chapters, and I'll discuss this use of what I will call "operational cyber" briefly here, but it's not the reason there is a separate cyber domain section of this book.

Here, I want to distinguish strategic coercive cyberwarfare from cyber operations conducted in support of conventional military operations and focus on this distinct class of more isolated and potentially independent cyber conflict. One can even think of strategic cyberwarfare as more than just a domain; it's a whole new way of using a certain type of force to coerce an adversary into desired behaviors, or to even force capitulation. The best analogy may be to the way strategic air campaigns were thought of by some theorists in the 1920s and '30s as decisive means, on their own, to compel an adversary's behavior. The idea then was that wars could be won by strategic air campaigns alone, using conventional bombing to effectively bludgeon a nation into capitulation by destroying its infrastructure and the ability of its economy to support conventional warfighting and society.[2] Conflict of this type in the strategic cyber domain can be either independent of or combined with conflict in more traditional domains. I have a hard time conceiving of a conventional conflict between major powers in which strategic cyberwarfare is not part of the equation—by both sides. Before I jump into that concept, let me start with a few words about cyber in support of future (or even current) conventional military operations.

Like EW, operational cyber operations will be ubiquitous and cut across all traditional military domains. I've covered that aspect of cyber as a complexity in the other domains. Every weapon system, all C3 systems, and all logistics support systems, as well as the commercial networks and

---

2   Although it has been disputed, one could characterize the NATO air campaign
     against Yugoslavia in 1999 as a successful example of a bombing campaign
     alone forcing a state to capitulate. The degree of dominance NATO had
     in this campaign was completely overwhelming. The American-Israeli
     attack on Iran is another attempt to achieve coercive aims by air power
     alone. As I write this, the outcome of this attempt is not yet known.

nodes they connect to, will have to be designed for cybersecurity to be effective in wartime in the context of conducting and supporting conventional military operations.[3] Legacy systems, which have been in the inventory for decades in some cases, are often the most vulnerable to cyber defeat. Military cyberspace offensive systems will be developed and employed as part of the spectrum of military capabilities any advanced nation-state fields. Conventional military or operational cyber has some characteristics that are important to recognize and understand and to distinguish them from what I'm calling the strategic cyber domain.

To be effective, operational cyber should be used in close conjunction with conventional military operations. Some degree of destructiveness can be achieved with cyberattacks alone, such as taking control of and de-orbiting an adversary's satellites, or in taking over a UAV's flight controls and causing it to crash, or even in redirecting a weapon to attack its own forces. However, most conventional cyber military operations, like EW, will be used to degrade an enemy's weapon systems or C3 systems to enable a more successful conventional kinetic attack. Cyber vulnerabilities are fragile; they can generally be readily corrected once they are known, and a successful cyberattack creates only a transient advantage. In general, attacking cyber military vulnerabilities buys the attacker some finite amount of time. Coordination with kinetic operations must be timed to exploit that transient advantage. Cyber vulnerabilities can be exploited through a synchronized conventional kinetic attack of some kind. That attack must be conducted while the cyberattack effects have not been corrected.

Operational cyber advantages are also unreliable. An adversary may identify and correct an existing cyber vulnerability at any time, the attacker's access may be cut off, or deception may be used to create the false

---

3   The last of the seven Operational Imperatives that I used to focus analysis of
     Department of the Air Force (DAF) modernization needs and opportunities
     was focused on our dependencies, especially in information systems, as we
     mobilized for and supported power projection operations against a peer. The
     DAF couldn't control all these dependencies, but we could at least expose
     deficiencies and make the case for additional investment to correct them.

belief a vulnerability exists where one does not. Tactical cyber advantages are opportunistic. They depend on the enemy having an exploitable weakness in one or more of its systems. The attacker must find and exploit those weaknesses where they happen to exist. The attacker may also not be able to determine with confidence the success or failure of a cyberattack, creating uncertainty and risk. All that said, cyber can certainly be decisive in support of conventional military operations, but relying upon it to be decisive or create an enduring advantage is a high-risk approach. In the conventional military domains, cyber will be an adjunct and complement to more conventional destructive military force.

The strategic cyber domain is concerned with using cyberattack, independently of other forms of warfare, as a coercive mechanism in its own right. This idea isn't entirely new. During the interwar years of the 1920s and '30s, there was a vigorous debate about strategic bombing campaigns, their potential to break the will of an adversary or their ability to efficiently destroy an economy's capacity to support both a population and a military enterprise. If successful, proponents asserted, strategic bombing could eliminate the need for traditional ground campaigns and even dramatically reduce the destructiveness, duration, and cost of war itself.[4] That debate is still active in some circles. The strategic bombing air-warfare theory was tested in the World War II air campaign in Europe in particular, with mixed results at best. The conventional strategic bombing campaign against Japan would not have obviated the need for an invasion and was decisive only with the introduction of nuclear weapons. The strategic air campaign idea was that a nation could be coerced by the physical destruction of economic networks, including manufacturing and transportation networks interior to the state, forcing capitulation or overthrow

---

4   Malcolm Gladwell, *The Bomber Mafia: A Dream, a Temptation, and the Longest Night of the Second World War* (New York: Little, Brown and Company, 2021) captures this debate and how it resulted in the strategic precision bombing campaigns against Germany and Japan in World War II. Notably, even before the introduction of nuclear weapons, both campaigns ended in massive casualty-producing attacks on population centers, as opposed to the limited precise and high-leverage strategic targets theorists had envisioned as decisive.

of the sitting government. The very same target sets, and more, are susceptible to a strategic cyberattack.

A lot of what I've written in this book addresses the potential to remove human beings from direct combat through lethal autonomous systems. Strategic cyber offensive operations finesse even that development, and possibly even the need for offensive conventional warfare itself, by putting populations and their governments directly at risk of catastrophic collapse through disruption or manipulation of their digital systems. I can't predict if strategic use of cyber would be successful in the future, but I can predict that it will be attempted; it's already being used by our adversaries.[5] Like operational cyber, it will also be used in conjunction with conventional military operations as well as independently.

Modern societies and governments are almost completely dependent on the flow of information through digital networks and on the processing and storage nodes they connect. Most wealth today exists only in digital form. Technology, innovation in business concepts, and economic imperatives are going to continue the movement toward an ever more digitally integrated and interconnected world. The recent introduction of massively scaled AI technologies, such as large language models (LLMs), and increased automation across society are only going to increase these dependencies. For the foreseeable future, there will be a profit-driven bias toward improved services, products, and efficiency over security. Security in the form of resilience has certainly improved, but absent some compelling event, cybersecurity will always be playing "catch-up."[6] For a nation prepared to make the investment in offensive strategic cyber, this presents

---

5   Scott Pelley, Aaron Weisz, and Ian Flickinger, "China's Capacity to
     Hack the US Is Growing, Former NSA Head Says. Here's What They're
     Targeting and Why," *CBS News*, October 12, 2025, https://www.cbsnews.
     com/news/how-china-targets-us-systems-tim-haugh-60-minutes/.

6   Richard Clarke, who years ago sounded the alarm about cybersecurity
     vulnerabilities, has become more optimistic in recent years as cybersecurity
     as a discipline and cybersecurity products have improved. Richard A. Clarke
     and Robert Knake, *Cyber War: The Next Threat to National Security and What
     to Do About It* (New York: Ecco, 2011); and Richard A. Clarke and Robert
     Knake, *The Fifth Domain: Defending Our Country, Our Companies, and*

an opportunity to create a very powerful, and unprecedented, coercive capability. We are already seeing cyber disinformation campaigns designed to influence the United States, weaken the government's legitimacy, and amplify and distort internal divisions. We are also seeing extensive theft of data, especially intellectual property and personal information (financial and otherwise). There have been reports of malware found in critical infrastructure, such as power grid control systems, and ransomware has been used to temporarily shut infrastructure systems down.[7] The capability I'm describing here goes beyond, but not far beyond, what we have seen or experienced so far; it provides the capacity to overtly compel a foreign state to affect a desired behavior through either cyberattacks or cyber blackmail, independent of other means of coercion.

How would one use this capability in practice? I doubt it could be used to force a nation to capitulate completely, allowing occupation and regime change, for example. Nations will endure a great deal before reaching those points. The threat of nuclear retaliation, for a nuclear power at least, would also conceivably limit the attacker's objectives to ones the threatened state could accept. For example, strategic cyber could be used to prevent a response to an act of aggression against something that was not seen as a vital interest. It could also be used to force a nation-state to change its international behaviors, reduce its forward deployments, or adjust its international policies. Any use or credible threat of use of strategic cyber would likely trigger a response that would reduce vulnerability and increase counterattack capability. This provides some but only limited deterrence to use. The most attractive opportunities for cyber coercion will be situations in which the aggressor places a much higher value on its

---

Ourselves in the Age of Cyber Threats (New York: Penguin Books, 2020). My own conversations with cybersecurity experts suggest that Clarke is too optimistic.

7   Jen Easterly and Tom Fanning, "The Attack on Colonial Pipeline: What We've Learned and What We've Done Over the Past Two Years," Cybersecurity and Infrastructure Security Agency, May 7, 2023, https://www.cisa.gov/news-events/news/attack-colonial-pipeline-what-weve-learned-what-weve-done-over-past-two-years.

objective than the threatened state does. There are plenty of examples, but the reunification of Taiwan with China could be on that list.

One other important potential use of strategic cyber is by a less capable adversary seeking to deter a great power like the United States from an expected adverse action. This may be the most likely use of strategic cyber; there would be a strong upside and very little downside consequence, even if the attack failed. Strategic cyber in this context would be seen as an offset strategy against a state, like the United States, that has dominant conventional and nuclear capability. For states like Iran and North Korea, strategic cyber could be a powerful, and very cost-effective, deterrent to intervention by the US.

The advent of strategic cyber, just as a possibility, opens a theoretical can of worms, or if you will, a Pandora's box, that has to my knowledge not been explored adequately. There may be a body of academic work in this area, but there is no general public or political understanding of the implications of strategic cyber. I won't try to fill that gap here, other than to point out that this needs to be done. A great deal of intellectual firepower was expended in the 1950s on the task of understanding what nuclear weapons meant for international relations, future conflict, and the threat these weapons posed to humanity and how to avoid a catastrophic nuclear exchange or event. Even today, that effort isn't complete, as a recent work by my friend Bill Perry makes clear, but our understanding has moved forward dramatically since the 1950s.[8] I don't think we have made even a good start at understanding the implications of conflict in strategic cyberspace yet. We certainly haven't made the necessary investments to protect ourselves from such an attack at scale.

Unlike nuclear weapons, which have a clear "used" and "not used" defining boundary, strategic cyber conflict has a continuum of levels and no clear boundaries or "red lines," and no internationally agreed norms of behavior. Questions that should be addressed include:

---

8   William J. Perry and Tom Z. Collina, *The Button: The New Nuclear Arms Race and Presidential Power from Truman to Trump* (Dallas: BenBella Books, 2020).

- When is a cyberattack an act of war?
- How would conflict in the cyber domain couple to or trigger both conventional and nuclear conflict?
- Is there a mutually assured destruction equivalent for strategic cyber? Should there be?
- Is strategic cyber inherently unstable—favoring a first user substantially—and if so, what could be done about this?
- How might the laws of armed conflict apply; are there practical forms of legal constraint on behaviors? If so, could they be implemented and enforced through international law?
- Is there a meaningful distinction between espionage and attack preparation, or even attack itself?
- Is there a distinction between physical destruction and digital destruction that has any utility in reducing risks of conflict?
- What would the characteristics of escalation be and could they be controlled?

These topics and many others should be deeply explored and publicly debated and there should be some international consensus on them. For now, however, I'll confine myself to describing future strategic cyberweapons and how they could be used operationally. These could be weapons of the future, or they could already exist.

OPERATIONAL CONCEPT

A strategic cyber operation would attack a set of targets, networks, and nodes on networks of high value to the state attacked. They should be systems essential to the functioning of the attacked country, the permanent loss of which would be unacceptable to the threatened government and/or population. There are no shortages of potential target sets for a strategic cyberattack. Integrated digitally connected systems of air, sea, and land transportation, commercial power generation, financial transactions, wealth storage, consolidated data storage, and communication systems, including commercial media essential to societal control, are all potentially vulnerable. The objective is coercion without provoking a response that is counterproductive—a nuclear counterstrike, for example, or invasion of one's country. For that reason, scalable effects are useful, even essential. The ability to crank up the intensity or to back away by reversing the effects of an attack, and on both the degree and permanence of the damage, have operational value. A combination of serious "signaling" destruction levels, with the threat to escalate substantially, would be operationally valuable.

Unlike everything else I've discussed so far for other domains, where autonomy removes emotion from the equation, this form of future warfare is all about psychological impact on the enemy's population and leadership. The strategic cyber domain isn't about operational cost and effectiveness exchange ratios and winning on a conventional battlefield. The goal of the strategic cyber campaign is to force a desired behavior without triggering an extreme reaction, such as a nuclear strike or an extended conventional conflict. If, however, a general or regional war is underway, strategic cyber can complement conventional military actions to help dictate the outcome of that conflict.[9]

It's worth highlighting that because strategic cyber is all about influencing humans, some forms of artificial intelligence play an important role in strategic cyber. An enormous amount of research on AI is about defining how to influence human beings, mostly to get them to buy things, but also to persuade people to accept and internalize political messages. A strategic cyber campaign is as much an information operation as it is a technical attack on digital systems.

A coercive strategic cyberattack, as I would imagine it, starts with the communication of three critical messages: (1) This is what we want, (2) This is what we can and will do to you if we don't get it, (3) Resistance is counterproductive and futile. It's simple and straightforward to commu-

---

9   The will of a population to endure attacks has often been underestimated. In many cases, even protracted bombing has only stiffened the resolve of the targeted population. That can certainly happen with strategic cyber as well.

nicate the first message. If an adversary is applying strategic cyber against the United States, it could be reunification of Taiwan with mainland China, withdrawal of air and naval forces from China's sphere of influence in the Western Pacific, the reabsorption of the Baltic States and Ukraine into Russia, or the withdrawal of US military presence from Europe. The second and third messages are communicated by words and actions. The scale and visible impact of these actions and communications are critical to success.[10] They must convincingly demonstrate both capabilities and intent.

A strategic cyberattack must therefore be tailored to the vulnerabilities and culture of the target. For a coercive strategic cyberattack to be successful, it must be perceived as overwhelming, and it must put at risk things of high, even existential, value to the defender. The psychological impact of the attacks on the defending government and population is the key to success, and this should be carefully analyzed by an attacker. A country's vulnerabilities to cyberattack are likely to be unique to that nation. Both culture and infrastructure characteristics matter. Totalitarian governments like China are very concerned about their control of the flow of information to their populations. Other nations, like the United States, may be more vulnerable to specific material disruptions. Crippling any of the various networks mentioned above would have a devastating impact on the population of the United States. When public will to resist is the target, attacks should have visible and visceral impacts on as large a fraction of the population as possible. The attack must be "real" and penetrate the consciousness of the general public at scale. In the US, public reactions would be swift for an attack on any of the critical infrastructure networks.

One national system that affects virtually every American continuously is the commercial power grid. Taking control of that grid and demonstrat-

---

10   The recent COVID experience in the US demonstrates the importance of communication to human behavior. Something as straightforward as wearing masks or getting a shot could not be communicated effectively to the American people. The fact that many people did not personally directly experience the impact of COVID and rejected information sources they didn't trust has powerful implications for how to structure a successful strategic cyber campaign.

ing the power to shut it down, including some physical destruction, would have a strong impact that could touch every American. Strategic cyber campaign planners don't have to guess about what would be effective. It's even possible to conduct covert experiments on a target population to acquire data to help AI analytics model with the target state's responses.[11] If that's considered too risky, data on a given society or culture can still be collected by observing the impact of random criminal and nuisance attacks.

In structuring a strategic cyber campaign, it would be most effective to attack several critical infrastructure networks and systems simultaneously. This would mitigate the risk of a failure in one or more targeted networks and have the highest motivational impact. It would substantially impact the attacked nation's ability to respond, and sow confusion. If possible, the attacks should be reversible and scalable so relief from the damage can be offered as an inducement, and gradations in severity can be managed to manipulate the targeted society. The more power and control the attacker demonstrates, and the more clearly it communicates there is no easy remedy or fix, the more likely the defender will accept that resistance and escalation are not in its interest. A coercive attack would also be more effective if the potential for more extreme damage, not yet inflicted but clearly available, was apparent and credible.

Simultaneously with launching attacks on an opponent's infrastructure, the attacker would need to take steps to harden its own critical networks and functions against a response in kind. Instituting a strong national cybersecurity program in advance and having the ability to sever, at least temporarily, all connections with extra-national networks, would be part of this concept.

Effective intelligence and counterintelligence are essential to preparation for and defense against a strategic cyberattack. Intelligence depends on network or system access and so does the ability to design and execute an attack. The concealment of the offensive elements of the concept and all the actions to prepare for execution is also essential. Given knowledge

---

11   Which makes one wonder if the ransomware attacks of a few years ago on oil
       distribution and meat packing in the US had an even more insidious purpose.

and time, any prospective attack can be defeated if it is discovered before it is executed. Deniability if attack preparation is discovered is also an attractive design feature. Most defensive measures for counterintelligence could be carried out as acknowledged cyber defense best practices, but some measures would also need to be concealed.

Offensive capabilities are only half of this operational concept. It may be even more important, especially for the United States, to define the defensive operational approach. How should a nation like the US protect its critical information infrastructure in the future? This can't be left to the marketplace. The marketplace is too fragmented, and the economic incentives do not motivate businesses and local or even federal governments to invest heavily in cybersecurity. I don't see any alternative to mandating through regulation that all critical infrastructure maintains high standards of cybersecurity. The government should also invest in and define the tools and network architectures, applications, and hardware devices that should be used to provide security and keep pace with the threats. Commercial hardware and software products that may contain cyberthreats or do not meet security standards must be excluded from use in critical applications, if not entirely.

The conventional wisdom is that the sophisticated cyberthreats will always be a step ahead. That may be true, but I don't accept it as a given. I'm certain the threat will always be well ahead if we don't invest adequately in the various aspects of cybersecurity and require their implementation. In the world of defense against nuclear weapons, with which I have a lot of experience, there is no such thing as a perfect defense. Even so, every nuclear weapon that is successfully defeated matters, and the mere fact of a defense acts as a deterrent. So it is with strategic cyberattacks. Imperfection is not a justification for inaction.

BUILDING BLOCKS

To acquire an offensive strategic cyber capability, one would first have to build a dedicated cyber force or cyber command with this specific mission. Conventional echeloned military organizational models can be applied to the other future domains I have described, but with unlimited flexibility as to relationships and unit sizes as the norm. For strategic cyber, something entirely new is needed. The offensive mission of this new organization is to create and field a capability to put other nation-states' critical civil infrastructure at risk in a way that will compel desired behaviors. It should probably have a matrix organization, with one axis being the targeted states, and the other axis being technology and target set-type expertise. This new military capability is independent of conventional military operations or nuclear operations, so this organization should be independent of the organizations with those missions. This organization, and its mission, should probably be concealed as much as possible, probably with a deceptive cover story. I say probably, because there is some value in having one's adversaries concerned about one's capabilities in this area. The new strategic cyber organization would be responsible for acquiring and operating the following:

- Access systems that would provide long-term access to foreign networks and nodes would have to be acquired and sustained. It would be expected that over time, some access points would be discovered and closed or diverted into deceptive dead ends. Some would simply age out as technology evolves. Redundant access systems would be highly desirable. As new adversary systems are acquired, new access systems will be needed.
- Intelligence and battlefield preparation or shaping tools that penetrate and collect information about foreign networks and systems and that create the opportunities to embed weapons.
- Operational weapons that could be used to shut down, slow down, or erase enemy systems would have to be developed on a case-by-case basis tailored to the target set. Weapons with

wide capability through broadly used commercial targets, at any point in a commercial computing stack, are desirable. To the extent possible, reversible and scalable effects would be desired. Some tools could cause irreversible or permanent physical damage (a la Stuxnet).[12] There should be a large and comprehensive inventory of tools with continuously evolving capability against emerging or new technology as it is fielded.

- Planning tools, including tools to analyze the psychological and emotional impact on societies and leadership, would be necessary for campaign planning.
- Testing infrastructure that would include interaction with humans and human behavior modeling as well as technical testing.

The defensive side of the strategic cyber domain has its own building blocks as well. Many of them are mirror images or counterpart of the types of systems and tools needed for offensive strategic cyber. I would envision a layered structure for a national strategic cyber defense system. The last layers in this chain are the cybersecurity provisions under the control of the organizations operating on the internet, government or commercial. If we are going to have effective defenses, we also need to ensure the networks themselves have layers of defense and the ability to detect and respond to attacks. This can be accomplished through regulation of the industries involved or through greater direct government involvement. The Department of Homeland Security is an obvious focus for either approach, but I would guess the bias for the foreseeable future will be for very limited government involvement in commercial networks. Events may change that bias.[13] Finally, we need a layer focused

---

12 David Kushner, "The Real Story of Stuxnet," *IEEE Spectrum*, June 20, 2025, https://spectrum.ieee.org/the-real-story-of-stuxnet.

13 One such event might be the discovery of the Volt Typhoon China-origin cyber actor's presence on critical US systems. "People's Republic of China State-Sponsored Cyber Actor Living Off the Land to Evade Detection," Cybersecurity and Infrastructure Security Agency, May 24, 2023, https://www.

on the security of the development and production of the key software and devices we use on and in our critical networks.[14]

Whether we like it or not, for the foreseeable future, there is going to be an arms race in this area between attackers and defenders. For the US, I don't think we can rely completely on the marketplace to generate the defensive tools, architectures, applications, and devices needed, but we can supplement, complement, and motivate that marketplace to be more effective. Government regulation is a building block for defensive cyber capability, but investment is also required.[15] A strong dedicated organization, separate from but coordinating with the offensive organization, is needed. In the US, the current organizational structure dedicated to this mission, the Cybersecurity and Infrastructure Security Agency (CISA), formed in 2018, is a start, but it is inadequate and needs to be reconsidered in light of the need for a stronger government role and more sustained investment.[16] Internationally, nations who wish to lower the risk of this type of threat must work together to develop more effective defensive techniques and tools, as well as countermeasures that can be widely used. The Cyberspace Solarium Commission in 2020 addressed many of these

---

cisa.gov/news-events/cybersecurity-advisories/aa23-144a; and "PRC State-Sponsored Actors Compromise and Maintain Persistent Access to US Critical Infrastructure," Cybersecurity and Infrastructure Security Agency, February 27, 2024, https://www.cisa.gov/news-events/cybersecurity-advisories/aa24-038a.

14 For many years, the Department of Defense has invested in trusted sources of electronics for military systems. "Intel Awarded Up to $3B by the Biden-Harris Administration for Secure Enclave," Intel Newsroom, September 16, 2024, https://newsroom.intel.com/corporate/2024-intel-news.

15 About twenty-five years ago, I was on an Army Science Board study that looked at cybersecurity for financial institutions. The firms we talked to at the time were not investing heavily in cybersecurity. They were gambling that something bad would happen to another firm first, and that as a result, the government would pay for enhanced cybersecurity provisions. Much has changed since then, but the need for government regulation and investment remains.

16 Under the Trump administration, CISA's budget and staffing are being significantly reduced. David DiMolfetta, "Multiple CISA Divisions Targeted in Shutdown Layoffs, People Familiar Say," *Nextgov/FCW*, October 13, 2025, https://www.nextgov.com/cybersecurity/2025/10/multiple-cisa-divisions-targeted-shutdown-layoffs-people-familiar-say/408773/.

concerns and is a good starting point for outlining a more effective defensive cybersecurity posture for the future.[17]

COMPLEXITIES

Fundamental Operational Needs

Command, Control, Communications, and Battle Management: By its nature, strategic cyber operates on largely commercial networks and dedicated networks that interface with the commercial internet-connected world. C3BM for protection efforts for commercial networks, including network infrastructure, data, and functional resilience, will be largely a commercial firm responsibility, although hopefully within standards established by the government. Events may force a stronger role for the government in this function, although not necessarily for the Department of Defense. If so, that role will have to be integrated into a C3BM structure that incorporates intelligence, defense, and homeland security, just for starters. The counterattack mission necessitates a strong operational C3BM link between commercial network providers and military responders. By its nature, and due to the importance of time in defense and response, much of this interface functionality must be automated, but there must also be strong human control of responses to ensure those actions are justified and appropriate. In normal peacetime, the use of this function is episodic. There is a background of ongoing malicious activity punctuated by intermittent significant incidents. If a threat mounted the type of coordinated, comprehensive, and decisively coercive type of cyberattack described above, the US would require a much greater capacity and scale for assessing and managing a response than it has, to my knowledge, acquired. The type of C3BM capability needed to support a large-scale counterattack or a large-scale offensive strategic cyber operation such as

---

17   "Cyberspace Solarium Commission," https://www.solarium.gov/.

I've described could be acquired today with appropriate investments, if the decision were made to do so.

Logistical Support: To acquire and maintain the type of strategic cyber force I've described would require a continuous pipeline of new attack techniques, means of access, and monitoring methods as well as up-to-date attack planning and management tools. The world of cybersecurity and cyber-related technology is incredibly dynamic. Any vulnerability will be fixed quickly as soon as it is detected or exposed, and change due to new technologies happens very quickly. Without a robust supply pipeline of new techniques, an existing inventory of attack options will decay over time.

Multi-Domain Considerations: Strategic cyber operations will often be employed in conjunction with conventional military operations. There are good examples of how Russia, in particular, has linked cyber operations against commercial networks to military and gray zone actions. These include operations against Estonia, Georgia, and most recently, Ukraine. A combination of disinformation, denial of service, and disruption have been used in support of military or quasi-military action on the ground in each of these cases. If strategic cyber is used in direct coordination with larger-scale military operations, it can provide a very cost-effective alternative to kinetic attack for disabling transportation and logistics nodes, for example. Most militaries, including the US, are highly dependent on supporting commercial networks for acquisition of supplies, transportation, and personnel management. In most scenarios, the US has to flow people and material overseas on short timelines and then sustain those forces. All the networks and software systems for managing these functions are potentially lucrative and cost-effective targets. Looking at it from the opposite perspective, the same may be true of our adversaries. There is no need to wait for future technologies to address these risks and opportunities.

Resilient Basing and Operations: Strategic cyber operations headquarters and facilities are basically office buildings and cloud computing environment installations for data storage and processing. With the introduction of LLMs, there has been an explosion of investment in data and processing centers. In a future conflict, they would be high-priority targets for any conceivable type of attack. One means of providing resiliency is probably deception—making the centers that support the military indistinguishable from all the other similar AI associated facilities. A combination of deception and redundancy could be most effective, as well as reasonable levels of physical and cybersecurity. In any event, resiliency, for both installations and their functions, should be an important design consideration.

Tactical Mission Variations (Offensive and Defensive Cyber Operations, Surgical Strike, Raids, Interdiction, Misdirection, Hostage Taking/Ransomware, and Escalation Control): There isn't much limit to the types of options a strategic cyber suite of capabilities could provide to national leadership. All the types of attacks the commercial world and governments have already experienced are on the list, and several more. Again, this isn't a technology-dependent future; it exists now.[18] The questions about what capabilities to acquire are more about opportunism, policy, and priorities than they are about technology. A caveat is these options can't be instantly available; there has to be an often long and uncertain effort to acquire access, sustain access, and build tools or weapons to exploit any discovered vulnerabilities.

Target Identification, Access, and Exploitation of Vulnerabilities: Before one can develop or launch a cyberattack on a target, it is necessary to find that target and obtain digital access to it. Once inside a target network, it can take significant effort to find and exploit vulnerabilities. This can take years, and access is a very perishable asset once it's obtained. Sophisticated opponents will also take steps to deceive attackers into thinking they have access to a target when they have been lured into a "honeypot" virtual

---

18   This appears to have been confirmed by the use of cyber weapons in the raid to capture President Maduro. "Venezuela Strike Marks a Turning Point for US Cyber Warfare," *Politico*, January 7, 2026, https://www. politico.com/news/2026/01/07/venezuela-us-cyber-warfare-00713507.

deception.[19] A problem with preparing cyberattacks against a target is the more activity within the target's systems, the more likely discovery will be. There is a very strong tension between intelligence collection needs and cyberattack preparation that may include implanting weapons in an adversary's system where they could be discovered. This work is necessary, however, for the preparation of strategic cyberattack options.

Design Requirements and Considerations

Anti-Tamper: A cyberattacker generally relies upon concealment prior to an attack, and once an attack is executed, the weapon used is assumed to be compromised. This isn't necessarily true, and one can imagine cyberweapons whose functionality includes self-deletion, or some other technique, to prevent an adversary from reverse engineering the weapon in question.

Attribution: A major power trying to coerce a desired behavior through cyberattack or the threat of cyberattack wouldn't normally be concerned with attribution. It would be a virtue. However, this isn't true for many other cyber operations, on either side. Attacks or intelligence collection can originate from anywhere with an internet connection and be routed through a large number of intermediate global locations prior to arriving at the target system. The origins of an attack can be contrived to create false attribution for any number of purposes. Attribution is a serious problem for defenders who would like to preempt an attack or mount a counterattack and/or exact other consequences against an attacker.

Collateral Damage Avoidance: There are multiple examples of cyberweapons that have ended up harming the wrong target or simply gotten out of control and wreaked havoc once released into the massively connected cyber universe of the internet. Stuxnet is one example.[20] In a large-

---

19   This reportedly happened in the instance of a Russian attack in the Netherlands. Gintaras Radauskas, "Pro-Russian Hacktivists Tricked into Fake Attack," *Cybernews*, October 13, 2025, https://cybernews.com/cybercrime/russia-hackers-cyberattack-honeypot-water-facility/.

20   For a summary of the Stuxnet episode, see "How Did Stuxnet Work? Inside the World's First Cyberweapon," *ScienceInsights*. March 7, 2026. https://scienceinsights.

scale strategic cyberattack, it also isn't hard to imagine significant unintended consequences, including loss of life and property. There is even the threat a cyberweapon will be "captured" and used on the originating nation. None of this precludes the development or use of these weapons, but it implies care has to be taken to understand and limit their effects.

Confidence in Automated Behaviors: Strategic cyber operations, once initiated, will be highly automated by their nature. There must be reasonably high confidence in the net effects of the attack. If threats are used to motivate the opponent, then the attacks must be convincingly consistent with that intent. Full-scale testing isn't always possible, but laboratory testing of software in the loop is essential, and some limited field testing is highly desirable. Scaling may be straightforward or uncertain and risky, based on the targeted system and attack type. Simulations can be used to evaluate scaling performance. Whatever the collection of verification means, the confidence level in a strategic cyber campaign's automated behaviors should be at least as high as that associated with a conventional military campaign. This concern is compounded by the extraordinary pace at which AI is maturing and the degree to which AI can be tasked to complete complex tasks autonomously. One could imagine giving an AI agent a task like: "Create a cyberattack on country X that will achieve the following political result." This could be done now or in the near future, but how much confidence would we have in the product the AI produced and what risks would be accepting to use such a product?

Countermeasures: Both on the offensive and the defensive side, there are and will be continuing efforts to develop countermeasures and to anticipate and overcome them. Some of these will be as simple as disconnecting from the internet completely, and others will involve very complex defensive measures designed to isolate threats, limit damage, and speed recovery. There is a robust market for cybersecurity tools that can be acquired individually and in combination.

---

org/how-did-stuxnet-work-inside-the-worlds-first-cyberweapon/#google_vignette.

Deception: Strategic cyber has its roots in the world of espionage and counterespionage, where deception is the norm and can take many shapes. For the US at least, the military community tends to deemphasize deceptive measures; it's part of the doctrine, but not a prominent feature of operations or an area receiving significant resources. The intelligence community, in contrast, lives on deception, and we should expect it to be a prominent part of designing and operating strategic or tactical cyber systems for all purposes.

Default Behaviors—Loss of Contact/Control by Echelon: Generally, cyberweapons we deploy "behind enemy lines," so to speak, are likely to be cut off from our control early in a conflict. A range of default behaviors is possible, as long as the autonomous agent we have deployed still has some degree of control over itself and its environment. Designs should include appropriate provisions for loss of supervision or communication, just as with any autonomous platform.

Escalation Control: Strategic cyber provides some interesting potential for escalation control, or at least the attempt at it. It is certainly possible to construct an escalating and de-escalating series of steps an attacker can use to try to influence the decision-making of an opponent. This consideration applies to both sides of the equation. For the offense, it's a matter of trying to construct rungs on the ladder of severity and psychological impact that have some degree of predictability. As discussed earlier, AI can be important here, but actual behaviors under stress can be difficult to predict, particularly with strong cultural differences. The range of options potentially available to an attacker is almost unlimited, at least in theory. Couple this with a range of other non-cyber steps and it creates an even richer playing field. A lot can be done to try to control escalation without crossing the line of employing violence against human targets. Examples in the case of China attacking the US include graduated cyber-based disruption of both military and civilian activities (including steps like taking physical control of military or civilian satellites through cyber penetration), plus jamming or even directed-energy weapon attacks on key satellites. For the US, there are potential cyberattacks on the Chinese economy, on military support systems, and on the control the Chinese Communist Party leadership exercises over public access to information. From the defender's perspective, there are a range of actions that could be used to negate or minimize the impact of all these possible steps and show one's own resolve. None of these tools will be available in a crisis unless they are prepared ahead of time and included in the strategic cyber arsenal and cyber C3BM system for possible implementation.

Humans in the Concept: For strategic cyber defense, I would envision an extension of what we have today, which relies upon humans as commercial network administrators and corporate chief information security officers (CISOs) and cybersecurity officers (CSOs) and their associated staffs of information technology specialists.[21] Law enforcement and intelligence organizations play a significant role today, and the federal government has some limited capability through DHS and CISA. DOD's role, through Cyber Command, is largely limited to technical security assistance and counterstrikes against attackers, all of which involve human expertise and informed decision-making. Automation will play a major and increasing role, but it must be managed effectively by humans. For an effective defense against a well-planned and executed strategic cyberattack, much more will be needed in terms of integrated headquarters and human decision-making capacity on policy, regulation, and preparation. I believe the US has an organizational and a human capital need for more effective operators, as well as better technical tools, that isn't being met today.[22] The lack of this capability leaves the US with some unattractive kinetic options for deterring and responding to a large-scale cyberattack as our

---

21   These are relatively new C-suite positions, and an indication of
      how serious cybersecurity has become to commercial firms.

22   As the Secretary of the Air Force, I tried to improve the department's human capital
      in the cyber area. Given the relative newness of cyber operations, growing senior
      leaders with real expertise in this area is a challenge. One step I took for the Air
      Force was to establish warrant officer grades for cyber and IT technical specialists.
      Warrant officers, which are roughly grades between noncommissioned officers
      (sergeants) and commissioned officers, are common in other military services
      where specialized technical training and current knowledge in dynamic technical
      areas are of importance. The Air Force had eliminated warrant officers in 1958.

default posture. For strategic cyber offense, I would envision integrated multi-disciplinary human campaign planning teams for each potential target, with associated sets of AI-based analysis tools to evaluate vulnerabilities of potential targets and assess the impacts of various courses of actions. Given what we have already seen, it seems safe to assume our potential adversaries have already implemented something like this.

Interoperability: At this point, my take is that any attempt to build an offensive strategic cyber capability would be primarily US only, at least initially. My view is the risks of compromise currently outweigh the value of interoperability, except for very close partners. There may be specific threats and specific cyberweapons where we want to move cooperatively and in some interoperable way with one or more international partners, but those should be the exception rather than the norm. Defensive cyber is just the opposite. The US and its close partners should be working together closely on measures to defeat and mitigate the threats to our infrastructure and critical national assets. We should be sharing information about threats and protective measures and working on common security standards. There may be times when these two regimes are in tension, but those issues should be worked out on a case-by-case basis.

Legal Constraints: This is a complicated area where a range of governing laws spanning privacy, search and seizure, criminal, international, and laws of armed conflict all have potential implications for military operations. In many cases, the implications of existing law in the cyber domain are not well developed and certainly not uniformly applied.[23] To some degree, cyberspace is largely the Wild West, although the US has traditionally taken the legal constraints on law enforcement and intelligence organizations very seriously. There is a huge problem today in the inability to enforce existing laws regarding intellectual property, privacy, data theft, and even espionage. The ambiguity about the threshold for a

---

23   This issue has become a matter of public debate due to the disagreement between Anthropic and the Trump administration over restriction on the use of Anthropic's Claude AI model to surveil Americans. See Frank Kendall, "What Both Anthropic and the Pentagon Got Wrong," *New York Times*, February 27, 2026.

cyberattack to be considered an act of war is rightly or wrongly considered a virtue by many; a clear line would effectively provide a license for activities below that line and would put states in an uncomfortable box for actions above it. I expect the US will (and should) make a strong effort to comply with legal constraints on cyber activities. I don't expect this of our potential adversaries. As we plan strategic cyber operations, we will need to consider any widespread suffering imposed on civilian populations, which may be precluded by the laws of war. Absent any international consensus or agreement, we will have to decide for our own planning purposes what level of cyber operation constitutes an act of war, with its concomitant legal constraints.

Operational Planning, Rehearsal, and Dynamic Replanning: Any strategic cyber campaign should be carefully planned, tested in simulation as realistically as possible, and include features that permit rapid response to foreseeable and unforeseeable target changes or broader reactions. A secure development and testing environment, with as much realism and the ability to scale or emulate scale will be essential. Digital twins have become a popular concept recently; for strategic cyberweapons, a digital twin is needed for testing and performance verification and concurrent development during an operation. In some cases, limited rehearsals or carefully conducted tests may even occur on adversary networks.

Perishability: Cyberweapons are very perishable. As soon as their existence is discovered (or even if it is not), the weakness that led to their creation can be eliminated—almost always through software and/or procedural changes that can be developed and scaled quickly. Even before a weapon is discovered, the vulnerability it depends upon may be eliminated due to routine security updates, other unrelated changes, or obsolescence of the system hosting the vulnerability. As a result, cyberweapons must be continuously evaluated for their effectiveness and newer weapons constantly pursued.

Physical Security: The physical isolation and security of all future strategic cyber-related activities is a given. As mentioned elsewhere, strategic cyber-related personnel and facilities will be high-value targets for threat espionage, sabotage, and attack efforts of all types.

Psychological Warfare Implications: Strategic cyber is a form of psychological or information warfare in that it is designed to influence human behavior without widespread violence. The design of a specific cyber-weapon may be influenced by AI-aided psychological analysis of the targeted society and leadership. The impact of the choices of targets and the choices of effects overall on those targets are highly important. This is the whole basis for a strategic cyber campaign plan. We know some attacks only increase the will to resist. As our ability to understand human behavior improves through AI-based analysis, we should be able to project the impact of a strategic cyberattack much more effectively and precisely.[24] The structure and sequencing of an attack, and the accompanying messaging, as well as the specific content of an attack, are all important to success. A campaign can end as soon as the reality of "overwhelming force" and the "futility of resistance" are understood by the targeted society and/ or its leaders. That is a mental process that the strategic cyber campaign should be designed to achieve as effectively as possible.

Quantum Technology Implications: The implications for the cyber domain of future technologies that will employ quantum phenomenology are significant. These technologies may not have a major impact during the period of interest here, but it's still worth noting their potential. Nearest at hand and already available commercially is quantum key distribution. Further away from practical use are quantum computing, communications, and sensing. Quantum key distribution provides for highly secure encryption between cooperating entities. An attempt to acquire an encryption key transmitted this way is detectable and self-defeating.

---

24 Much of the AI research in the commercial world has one purpose—to understand how to reliably manipulate people to buy offered products. The data the commercial world wants, and what internet service providers, major tech platforms, and social media companies have, is the correlation between purchasing decisions and everything else about the consumer. This all existed before the introduction of LLMs, which have massively increased the capacity to analyze and draw inferences from data.

Quantum computing will provide a tool that will allow for much more efficient code breaking against many widely used forms of encryption, as well as other specialized applications. Quantum sensing and communications will provide enormous improvements in sensitivity, precision, and data transmission. It's impossible to predict when each of these applications will be available for general use; some are being experimented with at scale today and others are in early or middle stages of development.

Reliability: Because cyberweapons are basically software, they can have classical software reliability or quality issues and can follow software reliability growth models. Cyberweapons tend to be single-use expendable products. Once a cyber "projectile" is fired, one can expect any opponent to react by quickly closing the metaphorical door the projectile flew through, so the same attack won't succeed again. Future cyberweapons will also operate on threat networks, where there is very limited opportunity, if any, to "burn in" the software and eliminate defects. Achieving high reliability, so that the first and perhaps only field use of a cyberweapon is successful, is a design requirement and challenge. As the use of AI to generate complex code continues to increase, assessing the reliability of that code will be a major challenge.

Requirements Creep: This is less of a consideration in the cyber domain than in others, but it still applies. Conscious attention should be placed on avoiding cost imposing marginal or low return-on-investment requirements.

Resilience: The concept of resilience—the ability to absorb an attack, continue functioning, and/or recover quickly—applies much more to defensive measures than to offensive cyberweapons. The continuing evolution of commercial cybersecurity measures has largely incorporated this concept into existing IT system and application designs. I expect this to continue. Resilience for offensive systems is largely about the ability to avoid detection over time, both for access opportunities and for weapons, once a weapon has been implanted in a threat system host.

Responsive Threats: The cyber domain is aggressively two-sided and dynamic. Any new IT system or application will be probed for weaknesses immediately upon fielding, if not in development. Efforts to close any vulnerabilities will often be reactive rather than proactive, but they will also happen quickly. Early in the internet era, the advantage was very strongly with the sophisticated attacker. That hasn't changed, but the size of the gap seems to have diminished as defenses have improved. The commercial world has largely accepted some level of cyberthreat risk because the economic incentives to do otherwise have been weak. Hardening critical national IT networks against the most sophisticated threats may improve in the future, but we can't expect to outpace the threat, especially the so-called "advanced persistent threats," meaning nation-states with well-funded, professional, and ongoing malicious cyber campaigns. Absent a breakthrough defensive technology event, which I don't foresee, we are just going to have to live in this world and design in as much hardness, redundancy, damage limitation, and graceful degradation capability as we can afford.

Single Points of Failure: Unfortunately, there are a lot of them, especially on the defensive side of the strategic cyber equation. Unless the government steps in and mandates designs that do not have single points of failure, we are likely to have them in our commercial infrastructure networks and systems. On the offensive side, any future strategic offensive cyber campaign design should strive to have multiple ways to attack any given target and a redundant set of targets as well.

Sustainment: Once a strategic cyber suite of offensive weapons has been integrated into a campaign plan, there will need to be an ongoing activity to sustain the viability of that plan over time. This implies several actions. First, the suite of weapons must remain concealed. Some cyberweapons may be deployed in advance and hidden on enemy systems (something to be avoided, if possible, because of the risk of detection and exploitation), while others may be deployed just before or at the outset of a campaign. In either case, the existence of the weapons and their means of access must be concealed over time. Next, the viability of any cyberweapon that is part of a strategic cyber campaign must be continuously assessed as a threat changes through upgrades and defensive measure adjustments. Finally, new weapons have to be in development continuously as threat targets change with the introduction of new systems and capabilities.

Test and Evaluation: It's largely impractical to conduct either offensive or defensive strategic cyber domain testing at scale, given the potential scope of a strategic cyber campaign. As a result, we will have to rely on a mix of individual system tests, selective extensions of individual systems, simulation, and higher-level modeling. Some defensive testing at reasonable scale is possible, especially for acknowledged commercial products. Some defensive products may (should) not be publicly acknowledged, however. Any offensive cyberweapon testing at scale would have to be conducted covertly. This is challenging and risky, but not impossible. Test programs will have to be developed to achieve adequate knowledge of offensive or defensive cyber system performance.

Training: During the Obama administration, we created a Cyber Mission Force to substantially expand the size of the Defense Department's cyber human capital account.[25] A lot of effort has been put into training those individuals. At this point, I would expect we have a reasonably well-trained group of people to defend military networks and for assistance in the defense of civil networks on an individual basis. I can't speak in any detail about our offensive capabilities. My guess is we have a long way to go in terms of preparing humans throughout the chain of command for their role in a large-scale defensive or offensive cyber operation of the type I imagine for the not too distant future. Before we can train operators, we have to develop the integrated C3BM system they would use and the suites of tools they would employ or deploy in a strategic cyber operation. The human machine interfaces associated with all of this should be developed in collaboration between those cyber warriors and the software engineers developing the needed weapons and control systems; in many cases, these may be the same individuals.

---

25   "Cyber 101: Cyber Mission Force," US Cyber Command, https://www.cybercom.
     mil/Media/News/Article/3206393/cyber-101-cyber-mission-force/.

Other Needed Military Functions

Civil Support: A great deal of malicious activity in cyberspace is not military and does not involve nation-state actors. The military and intelligence communities both have a role to play in assisting civil authorities to combat crime, malicious influence campaigns, and extremism, and in supporting efforts to protect privacy and intellectual property. There is a tension between these goals and the role of the military and intelligence communities in US society. My expectation is future threats will create pressure to overcome that tension for the common good. I can't predict how much that might change the defense establishment's role, but support of these functions is needed and should expand.

Electronic Warfare Support: There has been a lot of talk about the convergence of the EW and cyber areas. There is a relationship, but it is not particularly strong. Future strategic cyber concept designers should clearly consider all ways of accessing networks from both a defensive and an offensive perspective. Any RF portal (think antenna) can be a portal for EW applications or for cyber applications, at least in theory. In the commercial world, one can jam or spoof wireless networks and systems like GPS, as well as use cyber tools to protect or attack through these portals. Intelligence products of all types should also be shared without regard to the phenomenology involved in collection. Where it makes technical and operational sense, cyber and EW should be integrated, for both strategic and tactical applications.[26]

Intelligence and Counterintelligence: For a strategic cyberthreat to be viable, the intelligence on which cyberweapons are based must be exquisite

---

26   The Ukraine conflict has been a testing and experimentation ground for both cyber and EW. Over time, both sides have made extensive use of EW and cyber techniques and learned how to respond quickly to developments by the other side. Lennart Maschmeyer, "Cyber Conflict and Subversion in the Russia-Ukraine War," *Lawfare*, June 11, 2024, https://www.lawfaremedia.org/article/cyber-conflict-in-the-russia-ukraine-war; and, arguing against the strategic utility of cyberweapons: Grace B. Mueller, Benjamin Jensen, Brandon Valeriano, Ryan C. Maness, and Jose M. Macias, "Cyber Operations During the Russo-Ukrainian War," Center for Strategic and International Studies, July 13, 2023, https://www.csis.org/analysis/cyber-operations-during-russo-ukrainian-war.

and constantly updated at a rate comparable to the rate at which changes occur in the target set. Some target sets may be fairly stable, while others will experience changes ranging from periodic "patches" or security updates to complete technology refresh and system modernization. A lot of civilian infrastructure is old and infrequently changed, both in the US and in the cases of potential opponents. A significant effort would have to be put into achieving and maintaining access, understanding threat networks, and keeping any implants hidden and effective if activated. The counterintelligence effort must include not just hardening against threats but active programs to penetrate threats and stay a step ahead of their plans and operations. We already live in a world in which a shadow or gray zone struggle for intelligence and counterintelligence is being conducted. I expect this struggle to only increase in intensity as new technologies are fielded.

Psychological Warfare: This was discussed earlier because of its implications for system design. Strategic cyberweapons can also be used for more traditional psychological warfare purposes. Other operations, be they psychological, kinetic, or nonlethal, can be structured and executed with strategic cyber operations to have the best collective impact possible. This all puts a premium on understanding the motivations, biases, and emotional pressure points of the intended targets, a very rich environment to apply some forms of AI. The ability to analyze and even experiment with human behavior through AI techniques, even in real time, will dramatically increase the potential for successful psychological warfare operations supported by strategic cyberweapons and techniques.

Special Operations: There can certainly be a strong connection between strategic cyber and special operations, but I would not assign the lead strategic cyber role to the special operations community. I would imagine a focused, dedicated organization and effort with specifically cyber-focused subordinate organizations as the appropriate way to organize for strategic cyber domain operations.[27] I'm inclined to see special operations as having

---

27 I don't have an opinion and am neutral on whether a cyber service is the best way to structure this new organization. It is worth considering if strategic

the same relationship to a strategic cyber operation that the United States Special Operations Command has to geographic regional commanders today. Special operations should be integrated into strategic cyber domain planning. Operational use of cyber, on the other hand, should be integrated into more traditional special operations tactical planning.

CONCLUSION

One has to be careful about projecting that new modes of warfare may be decisive. The strategic bombing community made this claim before World War II.[28] What followed were years of conventional warfare in which, at least until nuclear weapons were introduced, strategic bombing was important but not decisive. We've never experienced either an active conflict for control of space or the type of strategic cyber campaign described in this chapter. In the case of the space domain, the dependencies of the joint force on space systems argues for its significance in a major conflict. In a conflict over vital or existential interests, I don't think we can make the same case for the cyber domain as decisive. I do think we can be certain that cyberweapons will be used on civil and commercial infrastructure in any major conflict in the future. If we are adequately prepared and can respond quickly, a strategic cyber campaign may not be effective enough to compel capitulation. If we do not prepare, capitulation may be the only realistic option.

---

cyber becomes an area of warfare in itself. I would not create a cyber service for operational cyber, which has to be very tightly integrated with conventional operations in the other domains. Creating a new military service implies an enormous amount of overhead cost for supporting structure unless those costs are born elsewhere more efficiently. For the Space Force, which is about 1/30th the size of the Air Force, those support functions are provided by the Air Force and the Secretariat of the Department of the Air Force.

28   Malcolm Gladwell, *The Bomber Mafia: A Dream, a Temptation, and the Longest Night of the Second World War* (New York: Little, Brown and Company, 2021).

CHAPTER 8

JOINT, MULTI-DOMAIN, AND COMBINED OPERATIONS

Joint, multi-domain, and combined operations have been around for a very long time. They have always been challenging.[1] The synergies of integrating operations across services, domains, and with allies are being emphasized now, however, and will continue to be for the foreseeable future. Advanced sensing, communications, computing, and integrated command and control technologies all provide opportunities to improve warfighting efficiency, both within and across domains and across organizational boundaries. The first modern iteration of this idea was "airland battle" in the 1970s. In the 1990s, the label "network-centric warfare" became popular. Today, it is a combination of MDO and JADC2, which has achieved almost a theological following among senior military leaders. The premise is technology has advanced to the point where integration and analysis of all sources of data, from all domains, into a timely decision-making process will confer a substantial advantage. My own

---

1   A good example is recounted in the recently released Ken Burns documentary *The American Revolution*, with its depiction of the problems that had to be overcome to conduct the American and French land and sea campaign that led to the surrender of British General Lord Cornwallis at Yorktown in 1781.

view is there are advantages to be obtained by operationalizing command and control and AI-supported battle management technologies more effectively and broadly. But I think the hype may be overblown for three fundamental reasons: (1) specific payoffs from JADC2 have not been quantified or prioritized and may be less than expected, (2) the enemy gets a vote, and (3) the focus on ideal interconnectivity and information flow may cause overreach and failure and may also prevent the United States from pursuing more fundamental changes in operational concepts and systems that will have higher returns on investment.[2]

Regarding overreach and failure, there have been ambitious attempts before to merge information more fully to acquire operational advantage. The Single Integrated Air Picture program undertaken by the Air Force was one. Another was the Future Imagery Architecture program to integrate space-based imagery. One I personally worked on was the Army's Future Combat Systems program. All these efforts were ultimately canceled. I don't consider these failures to be dispositive about JADC2 or the idea that future forces will be much more integrated across domains—I think they will be more integrated—but I also think designing those architectures and investing thoughtfully in manageable increments with specific high-payoff operational improvement is the right approach. Chasing undefined aspirational goals has made Joint C3BM the worst-performing class of programs attempted by DOD.[3]

The first step needed is to address item (1) above and determine what information moved where and integrated with what other information will be of greatest operational utility. Once that is accomplished, the next

---

2   When I first wrote the report for DARPA in 2021, JADC2 was all the rage. It still is, but it has now been supplanted, to some degree, by enthusiasm for rapidly fielding cheap drones.

3   When I was Under Secretary of Defense for Acquisition, Technology and Logistics I published a series of annual reports containing extensive data about the DOD's acquisition programs. This was one of the data points exposed by that analysis. Under Secretary of Defense, Acquisition, Technology, and Logistics, *Performance of the Defense Acquisition System: 2016 Annual Report* (Washington, DC: Department of Defense, 2016), https://www.govexec.com/media/gbc/docs/pdfs_edit/performance-of-defense-acquisition-system-2016.pdf.

step is to pursue that capability, assuming it is feasible, and field a meaningful increment of improved performance. Here's where I would look for operational efficiency improvements: First, in each domain, we want to know where the enemy's most valuable assets are and what they are doing, and we want targeting quality data compatible with our delivery systems (launchers) and weapons (projectiles). Next, we want to make tactical and operational inferences or projections from that data to inform our own force distributions and movements and to plan and order coordinated strikes on the most important enemy assets, while simultaneously protecting our own forces and achieving favorable exchange ratios. All of this must happen as quickly and accurately as possible. Because we can't afford to do everything, and because some things are not even technically possible, it's extremely important to make good choices about how and where to invest. I'll offer some thoughts on this (subject to future analysis) shortly.

As one thinks about specific high-priority payoffs, one also needs to consider the enemy's reactions. A unique feature of military versus commercial communication systems is someone is actively trying to disrupt or shut down military systems, physically destroy them, or wrest control of them away. Given the peer competitors the Unites States faces, we have no alternative but to design for hostile jamming and cyber environments and to harden against or avoid targeting for physical attacks of various types. Our C3BM systems must be robust, both intra- and inter-domain.

With size can come vulnerability and complexity. Our multi-domain networks must be designed as scalable integrated robust wholes; a piecemeal approach is unlikely to be successful. One popular technique is to analyze "kill chains" (I'll credit my friend Paul Kaminski with coining this term of art and form of analysis) or high-priority "operational threads." Both are useful approaches and good preliminary steps, but the so-called minimum viable product for JADC2 will have to be designed for maximum operational impact by enabling important kill chains or operational threads at scale (one-off engagements and sniping are interesting but not enough) and to be survivable or at least resilient with graceful degradation against all expected attacks. The key to joint and combined multi-domain operational efficiency will be timely and accurate management of complex many-on-many engagements using resources that cross both domain and functional spaces.

Currently, a lot of work is going into demonstrating individual threads of connectivity between systems that haven't traditionally talked to each other. This is fine, but it's an early baby step toward the much more difficult goal of integrated and highly automated operational battle management at scale. I've seen too many cases of C3BM overreach that ended in failure. What I saw when I became the Secretary of the Air Force was an idealized vision on the one hand that may be over-promising, and uncoordinated specific demonstrations on the other hand that were not clearly tied to achieving a well-articulated or quantified operational benefit.[4] Let's look at where those benefits might be most significant. I'd propose several integrated architectures, all of which I believe should be pursued incrementally and simultaneously, starting with the definition of a minimum viable product in each case.

OPERATIONAL CONCEPTS

There are several operational constructs where sophisticated integration can have a high payoff. Here are a few.

Countering an Amphibious Assault in the Pacific: If China has military ambitions with regard to Taiwan or any other offshore objectives in its near abroad, then surface ships are essential assets for pursuing those objectives. China's whole Anti-Access/Area Denial (A2AD) program is designed to prevent US interference with China's actions close to its own

---

4   As the Secretary of the Air Force, I attempted to address the risks and challenges associated with achieving JADC2's aspirational goals by placing all Air Force and Space Force C3BM under one leader, Major General Luke Cropsey. Shortly before I left the DOD, Luke was made responsible for Combatant Command Joint C3BM modernization as well. Progress is being made, but we have a long way to go. I told Luke repeatedly that I was giving him the hardest job I had ever assigned to anyone.

shores.[5] Surface ships are relatively targetable, and the target speeds provide opportunities for engagement with over-the-horizon launched weapons from air, land, sea surface, and sea subsurface launchers. While space-based sensors would be part of this concept, both air and sea surface, and potentially sea subsurface sensors could also be employed and integrated. The projectiles (anti-ship missiles of various types primarily) will be most effective if times of arrival, target deconfliction, and tactics are coordinated independent of the launch source of the weapon. One can imagine this operational architecture being built up over time from the most cost-effective building blocks. A key element of this version of multi-domain operations would be the full integration and inclusion of regional allies with their own land-, air-, and sea-based weapon and ISR systems.

Defeating a Ground Cross-Border Invasion: Integration of the systems that could defeat an aggressive land movement intended to seize terrain in Europe, most likely of a NATO ally: space, air, and ground sensors would provide enemy force movement and operational intent information and support targeting, primarily for weapon delivery from air- and ground-based launchers. This concept is a direct descendant of the FOFA systems developed in the 1980s as a result of DARPA's Assault Breaker project. A modern version would integrate space-based systems as well as unmanned aircraft (both as sensor and weapon transporter/launchers) and autonomous ground vehicles.[6] It would also be essential to integrate

---

5   I once briefed then-National Security Advisor Susan Rice on China's military modernization program. The main point I made to her then was that the US was the dominant military power on the planet, but as one gets within one thousand miles of China, that situation begins to change.

6   When I was Deputy DDR&E (Director, Defense Research, and Engineering) for Tactical Warfare Programs in the late 1980s and early '90s, the FOFA basket of programs reported to me as a separate high-priority office. The looming potential for the FOFA programs to checkmate a Soviet mechanized invasion of Western Europe may have been a factor in the attempts at reform and the ultimate breakup of the Soviet Union. FOFA programs were at the heart of the dramatic victory in Desert Storm, which, in turn, was a major motivator for the Chinese A2AD suite of programs.

NATO allies into this concept and coordinate fires from all sources for maximum impact.

Defeating Coordinated Regional Air and Missile Attacks: This requires integration of defensive counter-air and ground- or sea-based air defense systems into a single multi-domain C3 system.[7] Again, space-based ISR plays a critical role, but timelines are much shorter than for ground or sea surface targets, and automation will be necessary to respond effectively to threat attacks at scale. This isn't really new conceptually, but I don't think we've ever solved this problem for cases that involve large numbers of coordinated attackers. This construct has value in both the Asian and European scenarios. As a former Army Air Defense officer who served in the HAWK belt of air defense systems in Germany under the command of the Fourth Allied Tactical Air Force, and someone who was closely involved in the First Gulf War and the attempt to glean lessons learned from that experience, I can say with some confidence that integrating air defense and friendly air operations isn't easy.[8] When the threat includes cruise and ballistic missiles, as well as enemy aircraft (manned or unmanned) and intense enemy jamming, this is a "wicked" problem. For defense of critical assets against conventional weapons at least, it isn't an intractable problem, but solving it at scale will mean a reliance on automation and decision-making with imperfect data as well as close integration with allies. NATO has labored for decades to develop an integrated air defense command and control system, with only marginal progress to show for it. An incremental approach is certainly called for here. One helpful feature of the air domain construct I described above is that the willingness to accept some degree of attrition of unmanned systems by friendly fire should make this problem much easier. There is a high, even

---

7   I won't address the Golden Dome concept of air and missile defense of the US against strategic (nuclear) threats in this book for several reasons, not the least of which is that I don't consider it feasible or affordable.

8   I commanded a HAWK air defense battery in West Germany in the 1970s. Ukraine is using HAWK today. I once told then-Secretary of Defense Lloyd Austin that if it didn't work out for me as the Secretary of the Air Force, I could always get a gig teaching the Ukrainians how to use HAWK.

insurmountable barrier in the US for the idea that some losses of Air Force pilots to friendly fire from ground-based air defenses might be acceptable; it is not. For areas in which only unmanned systems are operating, there should be more willingness to accept risk.[9]

Integrated Space Control and Space Control-Related Operations Focused on Controlling and Dominating the Space Domain: Space control campaigns must be designed to both protect US assets and defeat enemy counter-space and space systems across multiple domains. The fight for control of space is likely to be determinative in a future conflict, but it isn't just about the space domain itself. It must include the kinetic and non-kinetic targeting of ground-based C3BM of space offensive and defensive systems themselves and of efforts to protect the ground- or sea-based space control and surveillance systems and space launch systems. The high premium for being the "first actor," or to achieve surprise in space, dictates that space-focused joint and combined operations be operationally ready to respond on short notice (minutes or even seconds) to hostile acts and across multiple domains.

BUILDING BLOCKS

While C3BM systems to support joint and combined operations will each have some unique features because of the physical characteristics of the domains and operational constructs involved, there is also a "common core" operational technical architecture, at least conceptually, that can be

---

9   In the US, this wicked problem is compounded by the fact that the Missile
     Defense Agency (MDA) is only directly responsible for ballistic missile
     defense, which excludes cruise missiles and aircraft. During my tenure as
     Under Secretary of Defense for Acquisition, Technology, and Logistics, I
     assigned MDA with the role of chief systems engineer for integrated air and
     missile defense, but my sense is that this was not seriously implemented due
     to service reluctance to accept any outside direction that might have imposed
     costs. For Golden Dome, a new four-star command has been established to
     direct the program. A highly competent and professional officer, General
     Michael Guetlein, has been given that responsibility. I don't envy him.

applied in each of the cases described above and in other cases. There is a similar "generic" problem at the heart of each of the various multi-domain problems. One can think of this as a meta-C3BM system that functionally integrates data from specific sources, especially targeting data, and processes it to provide optimized targeting assignments and an efficient allocation of resources together with plans to employ those resources. In each case, this common core meta-system must have the capacity for continuous updates, some degree of tailorable human oversight, and high degrees of automation that produce reliable, timely results. Another feature of this meta-C3BM system will be its scalability, so it can be employed as the operational C3BM kernel at various operational levels and for multiple operational problems. One approach would be a core set of capabilities with application programming interfaces (APIs) that specify formats for inputs on sensor data, weapons characteristics, force deployment and employment guidance, and other factors.

A critical design feature for any implementation of the common core I've described is the number and type of assets that can practically be integrated into the minimum viable product for the specific joint multi-domain instantiation. Each of the several examples I provided for joint MDO will have its own "sweet spot" at some level of force integration where technology and operational utility intersect. One immediate goal should be to define what level of integration that represents and to break down the barriers to achieving it. We can then design operational implantation around groups of capability, each of which is sized at that level of joint multi-domain integration. Our organizational structures, where we are likely to put human executive control functions, should reflect that same level of operational integration.

While the description of this building block may sound straightforward, it is anything but.[10] As noted earlier, the most unsuccessful programs

---

10   I can picture a block diagram of the core C3BM concept that looks
       roughly the same for each operational context. ISR information flows in,
       decisions are made with human oversight, and orders to force echelons
       flow out. There is a feedback loop from operational forces and continuous

in DOD history are joint C3BM programs.[11] There are several reasons for this, but the most important two are that these systems are very complex due to the massive number of requirements they have to meet, and that they require sound timely decisions involving compromises among the joint participants to move forward. I believe the technology exists now to effect the type of system I have described, and it will only get better over time. The harder problem, for the US at least, is getting the parties involved to cooperate, getting timely technically sound decisions on the needed specific design features, and achieving cross-service and organizational implementation of those decisions.[12]

COMPLEXITIES

Fundamental Operational Needs

Command, Control, Communications, and Battle Management: C3BM is the brain and nervous system for joint and combined multi-domain operations. I discussed the common core idea for tactical force management earlier and the integrated multi-service, multi-domain asset management function at roughly the operational level above. In the future, we will also require an integrated and hardened multi-echelon C3BM system to control and support forces at all levels. While we can and should use cutting-edge commercial technologies in this system, I don't think those technologies will meet military needs unless the DOD makes significant investments in adding security and resilience for our military systems.

ISR input and episodic order generation. Within the controlled tactical force, there are automated integrated behaviors specific to that force.

11　Under Secretary of Defense, Acquisition, Technology, and Logistics, *Performance of the Defense Acquisition System*, 2016.

12　For a very insightful work on the difficulty of creating large government software programs; Jennifer Pahlka, *Recoding America: Why Government Is Failing in the Digital Age and How We Can Do Better* (New York: Metropolitan Books, 2023). As the Secretary of the Air Force, I made it mandatory reading for my senior leaders.

Ideally, DOD should be a step or two ahead of the commercial products for performance as well, but given the amounts being invested in the commercial world, that may not be possible.

Logistical Support: If we are going to be successful against great-power opponents with any future operational concepts, we must have a secure and operationally resilient logistics systems that support the joint and combined force overall. While the number of people in the force concepts I've described is greatly reduced, there are still plenty of humans in the concepts and they will have to be logistically supported with food, medical care, and so on. Consumables must be provided, including fuel in some form and munitions. In many cases, forward maintenance is reduced, and some attrition of unmanned systems is expected, but there will still be maintenance and extraction requirements, as well as mobility support requirements for various needs. Automation, ubiquitous sensing, and advanced data analytics will play major roles in improving the efficiency of future logistics support systems. These technologies can be developed and integrated at a pace driven largely by civilian investments. What concerns me is the reluctance to appreciate the extent to which these systems are targets for both cyber and kinetic attack by our adversaries. If we don't design and harden our entire logistics systems, including their commercial dependencies, for the likely threats, then we are creating a very visible Achilles' heel for our future force.

Mission Variations: A joint or combined force could be called upon to do almost anything. The US military provides a massive set of tools that are usually not involved in major conflicts and available to support the nation as needed. The list in my memory alone includes humanitarian relief in a natural disaster, establishing a blockade, assisting in responding to a pandemic, conducting a hostage rescue, eliminating chemical munitions at sea, raiding a terrorist hideout, rebuilding or building a less-developed nation's military, responding to cyberattacks, enforcing no-fly zones, counter-proliferation, peacekeeping, and counter-piracy.[13]

---

13   Most recently, the US military has been tasked to provide border security, support federal and urban law enforcement, and conduct counter-

I'll briefly discuss these and others below. There is also a range of tactical missions normally associated with military conflict and the use of force: counterinsurgency, raids, covert operations, limited air campaigns, and so on. All the forces I've described were conceived as ways to become more efficient in a conflict with a peer or near-peer competitor, but while that may be the most important capability we want, it certainly isn't the only one. Any future force design will have to consider the full range of mission variations the total force should be able to perform. This will undoubtedly lead to some specialized units and equipment, and in many cases, to more human-intensive organizations than the ones I've described. My view would be to start by assessing the capabilities of the concepts I've described to deal with the range of possible missions. The assessment should focus on the adequacy of the joint and combined multi-domain C3BM structure and logistics structures. Once that exercise is complete, the need for specialized units, people, and equipment can be determined and prioritized.

Multi-Domain Asset Management Levels: If we want to be able to fight in various contexts using truly multi-domain forces, we have to think of force or asset management as a multi-domain and multi-service problem. Current joint forces are organized from below the Joint Task Force level (nominally a multi-wing, multi-brigade, and possibly multiple-carrier strike group-size force) on down by service component commands. Jointness currently tends to stop at a high operational level. In the various domain and multi-domain concepts I've described, there is generally a tactical level point in the chain of command at which I've placed humans who are controlling operations of what can generally be thought of as a single-service force, but with inputs and dependencies on multi-domain assets—especially space assets. If multi-domain operations are to be a reality, there must be a capacity to manage cross-service assets efficiently and responsively to support multi-domain and domain-specific operations with multi-domain dependencies. I don't think this should happen

---

drug operations. I won't discuss those uses of our military here.

at the tactical echelon I describe in each concept. It may happen one layer up, but in many contexts, it may be two layers up. The assets of concern are not just DOD assets, they also include intelligence community assets and allied or partner assets. I don't think there is a technical barrier here that is insurmountable, but the US has never been good at this and in many cases has defaulted to single-service solutions and creative expedients. In wartime, motivated leaders find work-around ways to do their jobs as well as they can, often with human-intensive brute-force methods. We will not have an effective multi-domain force in the future unless we solve this problem, which, like other barriers, is more about culture and leadership than it is about technology.[14]

Planning and Rehearsal: This is an embedded function in the C3BM system, but it's worth thinking about independently. Classic planning happens simultaneously on different time horizons—call them short, mid, and long term. The differences traditionally are measured in days, at least at the operational level of force management. The future planning process should be much more dynamic, situation-dependent, responsive, and continuous. It should also be much more automated. Rehearsal will be embedded in the planning process as part of planning and course of action analysis at the level where humans will be engaged and interacting. Both planning and rehearsal should be through sophisticated simulation and with accelerated time. I envision the future force efficiencies that will allow us to dominate on a battlefield will occur at two levels. The first is the highly automated tactical engagement level, with multiple platforms engaged against multiple platforms, using automated behaviors and engagement decisions. The second level is here, at the echelon in the joint or combined fight where humans exercise executive decision-making supervision over multi-domain operational planning.

---

14   The CJADC2 effort is intended to address this issue. When I wrote the original DARPA report in 2021, I hadn't seen much substance to support the rhetoric. There has been some good progress since then, but there is still much to do.

Design Requirements and Considerations

Anti-Tamper: We will have to assume that almost any, but especially the forward echelon C3BM nodes, will fall into enemy hands and be reverse engineered. We will also have to ensure commercial products utilized by DOD securely protect sensitive information, even if they fall into enemy hands.

Civilian Engagement and Interaction: All our forward-deployed joint multi-domain forces will have to interact with local civilians and likely with civil authorities. Some of these interactions can be accomplished or navigated by autonomous systems or through remote virtual contact, but many will demand some degree of established trust and intuitive decision-making that only humans will be making for some time into the future. Automated translation devices, for both audio and print, should make these interactions more productive and efficient.

Collateral Damage Avoidance: As in the individual domains, joint and coalition multi-domain forces will need to manage collateral damage avoidance consistently with the laws of armed conflict and hopefully with far fewer errors than in recent history. C3BM systems and target detection systems will have to meet established thresholds for false positives and include a continuous machine learning feedback loop to improve their performance. Effective human supervision must be provided over a range of situations and operational tempos. That cannot mean every engagement has to be approved once conflict at scale is initiated, but the C3BM system must provide for meaningful and effective control and supervision over a full range of situations.

Confidence in Automated Behaviors: As one moves up in chain-of-command levels or echelons and increases the reliance on offboard cross-domain sensing, C3BM, and fires, the complexity and the potential for unforeseen errors increases. Within this more complex context it is harder to anchor confidence in extensive field testing or operational experience, making it important to collect as much data as possible from live training or actual conflict and to integrate that data into modeling and simulation designed to boost confidence in automated behaviors at scale. I would expect any future suite of interacting systems involved in multi-domain operations would be subject to continuing confidence-building analysis and human oversight in a machine learning, continuous-improvement environment.

Counterintelligence: As both the US and its peer or near-peer adversaries move toward greater reliance on autonomy and connected architectures, the value of intelligence, and therefore, of effective counterintelligence, increases. Joint and combined C3BM systems and the people who design, build, and operate them will be high-priority intelligence targets. We Americans are not adequately attuned to the degree to which our systems, and we as individuals, are targets of intelligence collection. Some forms of AI will give us opportunities to be more proactive in our own counterintelligence activities, but we will have to do so within the legal constraints placed on searches in the US.

Countermeasures: At this point, I don't see any obvious opportunities or need for specific cross-domain countermeasures against current classes of weapon systems. The types of countermeasures noted earlier in each domain are generally self-protection systems. Designers of future cross-domain concepts should be thinking about integrated resiliency and effectiveness. As an example, one possibility might be a ground-based air and missile defense system with an increment of capability to defeat a low earth orbit co-orbital ASAT.

Cultural Resistance: This may, in fact, be our biggest impediment to achieving joint and combined multi-domain systems and concepts of operation. It used to be an emphasized Air Force written doctrine that "only an airman knows how to employ airpower." The other services have equally strong views, whether they are written doctrine or not. There are valid reasons for a degree of protectiveness over service prerogatives and roles, but those attitudes also serve as barriers to improvement the US cannot afford in a great-power peer or near-peer contest. My own view is the US needs a cadre of people who are truly joint, and only joint. This is a controversial subject and provokes a lot of antibodies, but we can't expect officers whose careers will essentially always be controlled by senior officers of one service or a single community within a service to ever look very far beyond that service's or community's vested interests.

Cyber and EW: With the partial exception of the undersea domain, neither cyber nor EW will respect any spatial boundaries, and any radio frequency antenna or even optical aperture can conceivably be exploited for both. As stated earlier, I feel the convergence of cyber and EW is not as great as others might believe. Nevertheless, the RF world is increasingly software-defined (functionality once performed by analog devices is now digital), making cyber a lucrative way to affect signals and a potential way to insert malicious digital content.[15] In the multi-domain world, attack options and cross-domain cyber and EW hardening must be considered and included in all designs.

Deception: As mentioned earlier several times, deception is underrated and underutilized in the US. It can be a huge force multiplier, especially at higher joint and combined echelons. It can also be a major contributor to deterrence, although relying on bluffs about nonexistent capability is a very dangerous tactic and an even more dangerous strategy. Deception should be a serious and conscious decision at the joint and combined levels as investments in future multi-domain capabilities are analyzed and compared. Doing this well is tricky because of the tension between deterrence and operational capability; convincing an enemy we are weak where we are strong would weaken deterrence, but the perceived potential to have concealed capability where we appear weak could strengthen deterrence, unless the weakness is discovered. The simple formula to render deception unnecessary is to be strong everywhere and make sure the enemy knows that. The US may try to be strong everywhere, but it isn't clear we can afford or achieve strength in all domains or in all combinations. The next best option is to appear strong everywhere and be strong

---

15 Software-defined up to the point at which RF signals are transmitted or received. Digital representations of signals and waveforms still have to be converted to radio frequency energy to be transmitted, and from radio frequency energy into digital representations after they are received.

where it's the highest payoff for the investment. We may want to appear weak where we are strong to convince an adversary to invest elsewhere, or make investments we can easily defeat. One can go on with this convoluted logic for a long time, but the default to not attempt deception at scale could forgo high-payoff options for both deterrence and operational success. Active deception programs are possible in all domains, but cyber, space, and undersea—because it's relatively easy to conceal capability in these domains—offer some particularly interesting potential.

Default Behaviors—Loss of Contact/Control by Echelon: As we move up in scale to cross-domain joint and combined operations, the implications of loss of control over elements of the force or of classes of capability (think total loss of space-based ISR, for example) become even more significant. This argues for some degree of redundancy in all critical functions. We should not make a major part of the total force completely dependent on a capability that might be eliminated or disconnected by enemy action. I've generally posited a future force that depends on functionality from space in every other domain. There would have to be some degree of degraded backup capacity for all the support functions provided from space. Structural resiliency and default behaviors for when supervision or external support are lost should also be designed into the architecture.

Energy: The energy needs of the entire joint and combined multi-domain force should be considered as a whole as part of the design process, not just for each domain or platform type. The current force consumes vast amounts of petroleum-based fuels for mobility and supporting organization operations. Fuel has been a limiting factor operationally on many occasions. It will be even more so when large, fixed distribution and storage systems are susceptible to attack by a sophisticated peer competitor. I do foresee more efficient energy sources becoming available, but I don't foresee a way to eliminate having a substantial logistics support tail for stored energy and the necessity to distribute energy in some form to the operational force.

Humans in the Concept: If we are going to fight in an integrated multi-domain way in the future, we will have to redesign our command and control organizations. For example, instead of an Air Operations Center, we might need an Air/Land/Space Operations Center, which would also include strong cyber and EW elements. As noted above, I would envision this type of headquarters being one or perhaps two echelons above the domain-specific C3BM nodes where human executive supervision of single-domain-centric operations occurs. The location, size, and capabilities of these multi-domain C3BM nodes would be tailored depending on the total force assigned and the mission context—form follows function. I can foresee a cadre of operators with a very special set of skills working in these units. Automated data analytics, decision support tools, course-of-action generation and analysis would be utilized and continuously upgraded through machine learning and feedback. These nodes would have to be logistically supported, secure, and redundant. They could conceivably be distributed or disaggregated and connected virtually. See the discussion of organizational structures and training below for more on this topic.

Identification Friend or Foe: It would be desirable to have a universal Blue Force Tracker for all platforms in the entire joint and combined force across all domains and including allies.[16] It wouldn't have to be an interrogation-and-response system like classical IFF, but it would have to be reliable and precise. The maritime automatic identification system offers an analogy, as do the traffic monitoring apps we all have on our cell phones and the RF tags for tracking commercial containers and packages in transit. I offer, as a working hypothesis, that technology can support this requirement and do so in a contested environment in a way that doesn't compromise friendly asset locations to an enemy. Having a system like this would pay fratricide avoidance benefits, but it would also be a source of exquisite friendly force situation awareness and support

---

16   Blue Force Tracker originated at DARPA and was used in
     Iraq and Afghanistan to provide friendly force location data
     (primarily for vehicles) on digital maps at all echelons.

operational and logistics planning. As an initial design goal, I'd suggest starting with a combined air and land domain capability that could be used to support friendly force situation awareness and control, as well as individual autonomous engagement decisions. As in other areas, the biggest obstacle to fielding something like this across the services may be cultural, not technical.

Information Management: The advent of generative AI and the ongoing massive investment in data centers and processing have all happened since I wrote the original DARPA report in 2020 and 2021. At the joint and combined multi-domain level, it is not so hard for me to conceive that technical limits will not meaningfully constrain data processing and analysis functionality. As exciting as new data storage, communications, and processing may be, they won't be infinite at the operational level of warfare. Future military information architectures at the tactical level will still have to be designed around real-life constraints, and as I've noted elsewhere, I think there will be a need to keep the bulk of the relevant data storage and data processing near the operational edge of the networks. We will also need to design for graceful degradation as information network nodes are attacked, overloaded, or severed from the network. Commercial cloud architectures are the current rage, and with good reason, but their resilience in a wartime setting must be considered in a future multi-domain construct. Optimal data management architectures change every decade or so as technology advances unevenly across the relevant functions. The commercial world will drive this; the military must stay poised to be a fast follower and integrate military requirements into emerging new information management concepts.

Intelligence Fusion: Future joint and combined operational concepts have to include powerful data fusion engines that operationalize disparate intelligence data sources across domains. The design challenges are significant but not insurmountable. Over the last few years, thanks to emerging technology, there has been a lot of progress on the technical ability to merge and employ data. However, two nontechnical obstacles will also have to be overcome. First, the operational and intelligence communities will have to merge—there can't be a sequential intelligence analysis function before intelligence is integrated into operational planning. Second, the services and intelligence agencies will have to come together on a single integrated multi-domain information architecture in which all source intelligence can be fused and employed on operationally relevant time scales. Some progress in this area has already been made, but much more needs to be accomplished and neither of these tasks will be easy. Both are essential if we hope to integrate intelligence into operations effectively.

Interoperability: Most aspects of interoperability within the US structure (cross-service information management, intelligence fusion, and so on) have been discussed under other topics. If we are going to integrate our allies and partners into multi-domain joint and combined operations, we will have to proactively determine where the highest payoff opportunities exist and work to realize them. I'd suggest full or nearly full interoperability is possible with some traditionally close allies, but in other cases, we will want to be more selective. In terms of payoff, getting real-time or near real-time access to allied and threat dispositions and getting allied targeting quality threat track data so that the US could include those targets and allied assets in its operational planning and dynamic retargeting would be high on the list. This is true across domains.

Legal Constraints: The law of armed conflict applies even more at this level and may be harder to implement. Processes that satisfy these constraints must be included in the cross-domain architecture. There will be many cases where engaging launchers or projectiles and even highly automated units may not have firsthand data on target ID, especially for beyond-line-of-sight engagements, which one hopes would be the norm. The criteria for approving these engagements must satisfy international law norms against indiscriminate collateral engagement, attacks on protected humanitarian and cultural sites, and others. In addition, at this level, constraints like overflight rights, encroachment on neutral territory, and so on must be included in operational planning. Planners have to take all the relevant constraints into account today. Future systems should provide even more effective and timely highly automated ways to impose the required constraints.[17]

Operational Planning and Rehearsal: Linking joint and combined multi-domain C3BM to enable planning and rehearsal at whatever scale is needed is analogous to the movement in engineering to model-based systems engineering (MBSE), which links models across the design stovepipes and across development, production, and sustainment. Under an MBSE construct, many design configurations can be digitally defined and tested through simulation. This is an evolving capability, but the trend is clear, and it applies in the operational world as well. The JADC2 and various service counterparts are intended to provide both operational execution and planning capabilities. There is an enormous amount of work to be done to make this real, but it has started, and I don't see a technical showstopper. I do foresee issues with cultural stovepipes, reluctance to accept the cost to change to accommodate integration with others, and difficulties with integrating even close allied partners. All this argues for beginning with one or two high-payoff use cases (at scale) to improve the chances of success and provide a meaningful increment of capability as a starting point.[18]

Organizational Structures: Current organizational structures are too rigid and too single-service. As noted above, for a major operation, a Joint Task Force is normally formed today with service components as the major organizational elements. Within the component organizations, one finds traditional structures like wings and brigade combat teams, which are certainly tailorable and can be augmented but which are still

---

17   It's an understatement that the second Trump administration has deemphasized legal constraints. "Maximum lethality, not tepid legality" in Secretary Hegseth's words. My own view is that American values support the rule of law and we want to be in a position to credibly demand compliance by others. Therefore, we should not relax legal constraints and that new technology will allow us to be even more effective at ensuring compliance without a significant operational penalty.

18   The approved JADC2 Strategy lacks specific mission applications or operational goals. In my view, any plan to implement the strategy should be focused on achieving specific operational results.

essentially single-service. To get the full benefit of cross-domain operations in the future, we will need more tailored and mixed operational structures that are designed around coordinated actions to achieve specific operational missions. As a starting point, we should consider future structures where every echelon above the basic tactical fighting unit is joint and involves the integration of operations in multiple domains. A space, air, land construct and a space, air, sea surface construct would be two reasonable starting points. At the other end of the organizational spectrum, that of total force capabilities, the US should also consider a Global Operations Command that provides multi-domain resources and support functions to all geographically regional warfighting commands.[19]

Physical Security: Joint and combined multi-domain organizations for operations of any significant size will be complex networks with important command and control, transportation, and logistics nodes. A thinking adversary will be looking for weaknesses and points to attack by any means that will have the largest impact on the force's ability to function. Fixed and easily targetable assets, such as current Air Operations Centers, are unlikely to survive. Concealment and deception can play a role in avoiding some attacks, but there will also be a need to physically protect against a range of threats, including special operations, covert operations, and sabotage. Unmanned security systems and barriers can play a useful role in providing physical security, but complex situations and innovative threats can occur, driving an enduring need for human-controlled and -managed responses to the full range of possible threats.

Reliability: The networks that support joint and combined multi-domain operations must be reliable to function over time and under stress.

---

19   This is probably a whole other book, but the basic idea is the US could use its global interior lines to organize some assets on a global basis operationally. This would include space, transportation, and logistics assets, some cyber assets, some airborne ISR and EW assets, global strike assets, and needed C3BM. For a great-power conflict, or even a regional one, this command could swing either way in support of operations, in Asia or Europe for example, as needed. It could be either the supporting or supported command, depending on the situation. The US provides joint global services now but in a piecemeal way from multiple organizations.

Redundancy and graceful degradation should be designed into C3 networks, for all types of attacks. Classical reliability—the length of time between equipment failures while in use—has to be adequate to ensure operations will not be degraded due to reliability problems. Commercial electronics should be adequate to support much of this requirement, if they can also meet all needed military requirements, including secure and trusted supply chains. Software and data systems that enable C3BM must also be reliable under operational stress, including under cyber and EW attack.

Requirements Creep: I'll play this record one last time. It is particularly relevant in this chapter. Conscious attention should be placed on avoiding cost imposing marginal or low return-on-investment requirements. The natural tendency in all domains, and especially in committee-designed multi-domain applications, is to add more requirements to the design until the concept crashes from its own weight. A lot of the items on this list are good examples of potential requirements creep. The whole point of the concept is improved cost-exchange ratios over current systems, which implies discipline in avoiding nice-to-have features. The abysmal history of joint C3BM systems is due in large part to overly extensive and complex requirements.

Responsive Threats: I would expect the first line of response would be to try to identify and attempt to destroy critical joint and combined C3BM nodes at each echelon. I would also expect attempts to emulate the capability described here, beginning with appliqués to current organizations. The US has been very public for years about its intent to rely on integrated multi-domain operations and joint C3BM. I'm sure potential adversaries are already working hard collecting intelligence and analyzing how best to defeat any important US advantages that materialize, and on how to field a superior version before the US fields its systems. Some existing threat systems, such as long-range surface-attack conventional missiles or anti-satellite systems, could be easily adapted to target joint C3 nodes.

Single Points of Failure: It I had to pick one capability of concern now, it would be PNT systems, especially GPS. All the ability to fuse target tracks, merge identification data, develop common operating pictures, conduct adaptive EW, enable cognitive sensors, and plan complex multi-domain engagements or operations depends on reliable position and timing information, currently supplied chiefly by GPS. Some surveillance systems can effectively be single points of failure given their multi-domain support functionality. A future force must expect any single friendly asset or small group of assets can be identified, attacked, and destroyed—if the adversary is prepared to pay the price of achieving that goal. An integrated multi-domain force must be designed to deny that opportunity.

Stovepipes: The DOD is currently designed and staffed around stovepipes of operational expertise, which, in turn, have their own communities of people who share training, unit assignments, and career paths with each other. Generally, these stovepipes are associated with certain classes of equipment (such as surface ships, artillery, and jet fighters). None of these communities are inherently joint or combined or multi-domain. All of them have their own cultures and are proud and protective of their operational roles. Future forces, and the people who command them, will have to break this paradigm to be successful. As we move toward greater reliance on human executive control of operations conducted largely by formations of autonomous unmanned systems operating in a multi-domain environment, we should also create operators who are inherently multi-domain and who become skilled at joint and combined operations early in their careers.

Sustainment: Managing the sustainment of a joint multi-domain force that is highly dependent on autonomous unmanned systems will bring a set of unique challenges, not the least of which will be avoiding providing an adversary with high-payoff vulnerable targets. One risk that will have to be addressed is the potential dependency on cloud processing and data storage and the possibility that cloud infrastructure may provide "cheap kills" for an adversary, by direct attack or by attacking dependencies like the power distribution system. The entire supply chain will have to be resilient and structured with threats in mind. Special attention will have to be paid to having wartime consumables, including munitions and high-demand spare parts, and survivable transportation to get these and other items where they are needed. One should expect the range of possible attacks, from cyber and long-range kinetic precision weapons to sabotage, to be employed.

Test and Evaluation: Testing at scale and testing against likely responsive threats will be challenging, but both are necessary. A lot can be accomplished with linked virtual human-operated C3BM nodes and a mix of real and simulated platforms and autonomous platforms and units, but there will need to be a realistic set of data based on field data to anchor the virtual simulation world. The test, evaluation, training, and rehearsal worlds all have a lot in common, but there must be adequate resources to meet each need.

Track Fusion: This challenging problem is at the heart of realizing battlefield efficiencies from multi-domain integration. It enables the efficient allocation and coordination of strike forces and precision weapons. History suggests strongly this capability should be developed incrementally, starting with the highest-payoff use cases and growing from there. Single-platform fusion of multiple organic sensors should come first, followed by integration of those tracks with the highest-payoff independent multi-domain sensors. The technical issue isn't just making multiple sensors for a single-threat track merge, but correctly sorting out for engagement decisions and weapon allocation many tracks in a confused and cluttered many-on-many battlespace against an enemy that is doing its best to confuse us.

Training and Experimentation: A new class of joint operations military professionals, of all grades, will be needed to achieve the future multi-domain concept I've described. They won't be developed and trained in the existing service-specific professional military education systems. They will fight intensely, but from screens that control multi-domain forces. They will need to train close to continuously. They won't be adequately trained through infrequent exercises in which threats are curtailed to provide a more benign operating environment so all exercise goals can be met, something that is the historical norm, especially for problems like jamming and cyberthreats.

Other Needed Joint and Combined Military Functions

Air and Missile Defense: This function was described in the air domain and the sea surface section, but it is a concern in all domains and a major reason for linking domains together. Future space sensors must provide early warning and tracking adequate to hand over threats to other air- and ground-based sensors, and at some point, directly to weapon (projectile) seekers. Air and missile defense requirements cover the tactical, regional, and intercontinental threats and the range of ballistic and cruise missiles and aircraft, including hypersonic weapons. They also include warning and indications information and analysis and support to strike against ground targets, including mobile launchers. All this functionality, including that required for strategic (nuclear) defense, should be designed together to provide a balanced suite of multi-domain capabilities.[20]

Airlift and Sealift—Strategic and Tactical: The forces I've described have transportation needs of various types, but much less than current forces. Some forces can self-deploy, but all will need sustainment support. I haven't tried to size the dedicated military lift assets that would be required. A lot depends on the degree to which prepositioning is used to reduce deployment time. Given emerging and future threats, I would support keeping as much material forward as possible. As noted earlier in the book, ground combat units at scale, even with the lighter vehicles and reduced ground support fleet of trucks envisioned here, must come predominantly by sea. The mass and volume are too great to do otherwise. This makes the large ground force deployment lead time weeks instead of days or hours. By relying largely on UGVs and a combination of small and medium UAVs for the ground domain, it is relatively easy, from the per-

---

20 Golden Dome would appear to address this problem for population defense of the US against nuclear threats, but almost nothing has been revealed about the mission requirements or integrated concept for Golden Dome.

sonnel management perspective at least, to preposition ground domain equipment forward, reducing the need for sealift. The low numbers of humans in those forces also reduces the need for airlift of personnel. For both types of lift, unmanned systems are a possibility, but using manned systems does not impose a heavy penalty. As forces are sized for the relevant operational plans, air- and sealift can be sized to support those forces and plans.

Chemical and Biological Warfare: As chemical and biological threats are used largely to kill or incapacitate humans, the reduction in humans in the forces I've described should reduce the incentive to acquire and use these weapons. Nevertheless, they could still be used against specific targets, and any place where humans live and work will have to be protected. Most chemical and biological detection systems today require close proximity and physical sampling. If practical airborne sensors can be developed, that would be a valuable capability. In any event, the multi-domain C3BM system will need the ability to efficiently integrate chemical and biological warfare-related information and to use AI-based tools to analyze that data and recommend actions.

Close Air Support: This was covered in the Air Domain chapter, but it is a consideration for multi-domain forces. I've always been a big proponent of this capability as it is an important enabler and source of protection for troops in contact on the ground. With the combination of the reduction in humans involved in line-of-sight combat and the availability of large numbers of lethal UAVs, it isn't as significant for the air domain forces to provide dedicated support to troops in contact. As a result, airpower can be allocated with a higher net effectivity. That said, there is no reason why the envisioned air domain assets couldn't provide this function when asked to do so. It should be an inherent capability of the multi-domain force.

Combat Rescue: This was also covered in the Air Domain chapter and should probably be thought of more as emergency extraction, as there should be few cases of aircrews forced down behind enemy lines. I envision a small fleet of unmanned vertical-lift air vehicles that would be used in certain situations to extract people, and possibly also insert or extract equipment, in hostile territory. This could be common with the tactical vertical-lift capability described in the land domain section, but with more survivability features.

Counterinsurgency: It's possible the US could be drawn into another direct military role in a counterinsurgency campaign. This would certainly be a land operation with strong air, space, and cyber domain elements. Like counterterrorism, the US military should be in a supporting role for other elements of power and local military forces, but there would be a stronger military element. The use of unmanned systems would permit more restrictive rules of engagement (up to a point), something that would be beneficial to the goals of a future counterinsurgency campaign. In a persistent counterinsurgency situation, I would expect the ad hoc application of emerging technology to the situation's specific and unique characteristics, but hopefully with less lead time than in the recent experience of the US.

Counterterrorism: Counterterrorism is not fundamentally a military problem; it's more about law enforcement and public physical security. The future military will continue to play an important supporting role when military force needs to be applied. Most military counterterrorism operations would be conducted by joint and combined special operations elements. There is a logical extension of the extensive capabilities to conduct these sorts of operations (surgical strike, hostage rescue, raids, seizures, interdiction, and so on) that were developed after 9/11 but with the future integration of emerging advanced technologies. These operations inherently blend elements from multiple domains and are often conducted with friendly non-US forces. They also involve intelligence and law enforcement personnel from outside the defense establishment. Humans will continue to have major roles in these missions because of their complexity. New tools should be available to support these missions from a range of technologies. One aspect of counterterrorism that may advance dramatically is information operations, which may not be a military operation per se but could well be decisive at addressing root causes and the ability of terrorist groups to recruit and grow.

Disaster Relief and Humanitarian Assistance: The multi-domain capabilities of the US military will continue to be called upon to support this mission. These include space and airborne surveillance and situation awareness, emergency logistical support, medical support, engineering support, and evacuation. This won't be a major driver on force design or equipping, but there will still be applications of military capabilities to this mission. Some of the capabilities of the multi-domain forces I have described will enhance the ability to perform this mission.

Gray Area Influence Operations: Included under this rubric are information operations, sabotage, paramilitary operations, and population intimidation in various forms. The US military may be asked by an international partner to assist in the defense against these types of operations. The US should have the capacity to bring integrated multi-domain capabilities to the table, but I don't see this mission as a major driver of force design or equipment choices in general. A combination of this mission, counterterrorism, and counterinsurgency could lead to the inclusion of some specialized multi-domain units in the total force.

Medical or Noncombatant Evacuation: There will be some inherent capacity to provide this support function by air, land, and sea. Because the future joint force will be much less human-intense, it may not be close to the scale of capacity the current force has.

Mine Warfare: The multi-domain aspects of mine warfare largely involve use of multi-domain sensing to facilitate effective mine deployments, detection of mine emplacement activities, and support to operational mine clearance. Cross-domain requirements for sensors and information processing systems should include mine warfare-related needs and opportunities.

Pandemic Support: This is not a core mission for the joint force and won't be, but the inherent capabilities of the military can be used to support this mission. As with medical evacuation, the force will have a lower need for medical capacity relative to the current dense human force, so the inherent capacity may be much less than currently.

Special Operations: Some special operations forces missions have been addressed above under the counterterrorism, gray zone, and counterinsurgency topics, and/or in the other domain-specific Chapters. The US will continue to need a suite of multi-domain special operations capabilities and specialized units, for use in any type of mission. This is likely to remain a human-centric capability, with autonomy and specific forms of artificial intelligence applied to various aspects of the mission over time as technology matures.

Training and Assistance Missions: The joint force will still be asked to perform these missions, but its capacity to provide that support may be greatly reduced with the shift to unmanned systems. As our allies move in the same direction, there will be a need to train together for joint and combined operations using multi-domain capabilities. The ability to train less-developed nations with more traditional conventional forces will be curtailed relative to current capacity.

CONCLUSION

Achieving the high potential enabled by autonomous lethality and AI will require major changes at the joint and combined levels and a rethinking of some fundamentals about how forces are organized and conduct operations from the top down. The greatest obstacles to achieving this goal are institutional, not technical. If requirements can be limited to those needed to achieve substantial improvement in specific high payoff operational capabilities, and if organizational and cultural barriers can be overcome, then major improvements are within reach.

CHAPTER 9

SOME FINAL THOUGHTS

Warfare has a logic of its own. When new technologies emerge that enable more effective military forces, those technologies will be adapted to that purpose. Sometimes, new technologies affect the established identity of a "warrior class" whose existence and status are tied to the application of older technologies. When specific warfighting roles and identities are threatened, change can be resisted fiercely. But the change comes anyway. Military professionals whose status depends on their roles and who have been successful in the past are the most inclined to resist change.[1] In addition, success does not breed operational innovation. The conventional wisdom is that Germany's defeat in World War I led to the development and adoption of the blitzkrieg operational concept, even though other countries had similar or superior technology. The US has had extraordinary success in conventional operations since the end of the Cold War, at least against nation-state militaries. However, the US has not contended with a true technological peer competitor for over thirty years.[2]

---

1   After Crecy and Agincourt, mounted armored knights spent decades trying to improve armor and outlawed crossbows rather than accept the end of their status as elite warriors.

2   Today, the most senior leaders of our military have almost no experience with a peer competitor that is well-resourced and focused on fielding forces

Over the last decade, alarm bells have rung more and more loudly and insistently about the threat to US power projection posed primarily by Chinese and also by Russian military modernization. My voice helped start that process in 2010 when I returned to government after a fifteen-year absence. Later, as the Secretary of the Air Force, I never stopped emphasizing China, China, China, and focused on preparation for great-power conflict as the driver for every decision I made. Despite this growing recognition, I believe the US is still not gaining ground fast enough in the competition with China for conventional military superiority. We may not be prepared to embrace the degree of change that warfare's independent logic is going to impose on us. Change brings risk, but so does complacency and inertia.

Certainly, steps have been taken, going back to the Obama administration, to address Chinese and Russian modernization initiatives. However, I don't believe those steps are adequate. Many of them have been in the wrong direction or focused on the wrong problems. There has been too much emphasis on speed for the sake of speed. I'd like to think that I made significant progress with the Air Force and Space Force while I was the Secretary of the Air Force, but I know much more needs to be done.

Prior to the conflict in Ukraine, the US was taking three strategic approaches to respond to the emerging conventional force threat to our power projection capabilities. First, the US has sought alternative sources of technology, largely reaching out to the commercial start-up world symbolized by Silicon Valley. Second, the US has tried to accelerate the fielding of new systems through higher-risk "rapid" acquisition approaches. Third, the US has embraced a largely aspirational goal of integrating operations through advanced information and AI technologies. All these initiatives have positive attributes and can provide improvements in fielded capability, but at least as implemented, none have provided the kind of next-generation game-changing capability the US actually needs. In some

---

designed for the purpose of defeating the United States. I was a Cold Warrior for about twenty years. The current Chairman of the Joint Chiefs of Staff, General Dan Caine, was commissioned in 1990, just as the Cold War ended.

cases, they are making the problem worse. The outreach to commercial technology sources for innovation may accelerate the flow of some technology into the Defense Department inventory. Competition among defense contractors, however, already provides strong incentives for this to happen. Most recently there has been an obsession with speed and a push to proliferate and accelerate the fielding of inexpensive drones and to expand the defense industrial base.

The innovation that's missing from all of this is on the requirements side of the house—the demand side, not the supply side of the so-called "valley of death." Overly aggressive development schedules are more likely to lead to program failure, higher cost, longer schedules, and lower-quality products than they are to revolutionary capability fielded dramatically faster. There are decades of experience and data to support this statement. Integrated joint C2 efforts alone will not make current platforms and capabilities significantly more survivable. In my view, there is only one way to change the trajectory the US is currently on, and that is to determine what the next generation of conventional warfare will look like and to get there first. I don't believe it will look like improved versions of the things we already have, even if they are integrated more effectively. I also don't believe it will look exactly like what we are seeing in Ukraine so far. The conflict in Ukraine has made it clear that uncrewed systems, primarily UASs or drones, but also uncrewed vessels and ground vehicles, are changing warfare in fundamental ways. Recently, DOD guidance has been to accelerate the adoption of these technologies and field large numbers of uncrewed systems, primarily UASs, as quickly as possible. We need to give more thought to where we are headed. Ukraine has given us a pointer in the right direction, but it isn't the final answer.[3] Running faster in the wrong direction will not get us where we need to go; it just consumes our most precious resource—time.

---

3   For an excellent discussion of why what Ukraine has done under the enormous time and resource pressure it is facing doesn't represent an optimal course of action. See Vitaliy Goncharuk, "Ukraine Isn't the Model for Winning the Innovation War," *War on the Rocks*, August 12, 2025.

I am convinced that unmanned lethal autonomous systems, operating in conjunction with one another, and which have roughly the characteristics I have described in this book, will be that future generation. Over the months I spent originally writing the first report for DARPA, and the years I spent leading the Air Force and Space Force, the evidence this is where the future of warfare lies has only increased. I am equally certain I haven't got the specifics of those concepts right, but I hope that I have provided a reasonable starting place for thinking about, analyzing, modeling, and experimenting with these ideas and others. The US needs to get on with that work urgently. China is being very innovative as it modernizes and is not waiting for the US to field new capability to respond. Relative time to market is everything when one is engaged in a highly competitive enterprise.

If I'm somewhere near correct in the view that lethal autonomous systems will, to a large extent, replace humans and human-crewed systems in future conflicts, then there are some questions we should try to answer and some implications of those answers that we should consider:

- Will conflict initiation and conflict termination be fundamentally different?
- Can escalation be controlled and if so, how?
- Will the first-mover advantage be highly destabilizing?
- Would superiority in and acceptance of the risk of use of AI-aided and autonomous operational decision-making be decisive?
- Could either space or cyberspace alone be a decisive domain?
- How could the scope of a conflict in duration or extent be limited, if at all?
- What are the implications for proliferation of weapons of mass destruction or for the potential employment of those weapons, especially nuclear weapons?

These questions and many more are all driven by risks we are not currently even having serious discussions about. Some risks could provide powerful deterrents to future conflicts. These include escalation to nuclear war, economic disruption, loss of societal control by established elites, the potential for unforeseen catastrophic operational failure, or conversely, for a lengthy and industrialized war of attrition. But cutting the other way may be the potential perception, real or imagined, that geopolitical or other desired end-states can be easily achieved, likely by surprise, by initiating a short decisive conflict from a position of technological superiority.

Let's all hope these questions are moot, and the future of warfare is a null set. War is a stupid and wasteful mechanism for solving disputes among humans. Unfortunately, the alternatives require some willingness to accept a partial loss of control over decisions—acquiescence to the rule of law in some form. Over the centuries of human civilization, the rule of law has gradually expanded to replace personal vendettas, feuding between clans, violent conflicts between city-states, and most, but not all, tribal and religious warfare. There are many exceptions, of course, but the overall, trend has been in the direction of reduced violent conflict. Attempts to establish mechanisms to eliminate wars between nations have made some progress over the last century, but the specter of large-scale conventional, and even nuclear, conflict still haunts us all.

Autonomous-machine-based or not, the future of warfare will still depend on human decision-making about starting wars. Unfortunately, the raw material of human psychology and its manifestations in society's collective behaviors and in individual human leaders hasn't changed much over the handful of centuries of human organization in large societies. Where will the movement toward conflict occurring primarily between autonomous uncrewed machines instead of between human beings in crewed platforms that I anticipate in this book take us? Perhaps the folly and irrationality of using violence to resolve our differences will become clearer and more compelling. I fear, however, that the result will be the opposite.

ACKNOWLEDGMENTS

With thanks to Steve Walker and Peter Highnam of DARPA for giving me the original opportunity to put down these thoughts in a secure setting at DARPA, to Ed Oshiba from the Department of the Air Force for his help in making this book possible, and to the many people whose wisdom, knowledge, and experience contributed over the past half century to the ideas expressed herein. Special thanks to my sons, Scott, Eric, and James and to my wife Beth for their support while I did my best to contribute to the effort to keep our nation safe and support American values.